How To Draw Dinosaurs

Volume 1

By Tracy L. Ford

ISBN-13: 978-1511934824

ISBN-10:1511934824

Table of Contents

Dedication

I dedicate this book in memory of my good friend Stephen Andrew Czerkas. He was an exceptional artist and paleontologist. We had many an interesting conversation and I'm dedicating this book to him because he believed in me, and supported me and my research.

Acknowledgements

I would like to thank Mike Fredericks for letting me write the "How to Draw Dinosaurs" articles, and for his editing. In late 1995 (I think) I started buying Mike's magazine "Prehistoric Times" and bought the back issues at the time to fill out my collection (I'm happy to say I have a complete collection of Prehistoric Times). At the second Dinofest held in Arizona (1996) I had the pleasure of dining with Mike. It was then that I brought up the idea of starting a series of articles on how to better depict dinosaurs by using their anatomy. He was up for it and shortly after that I started to write the articles for each issue of the magazine. I have fun writing the articles and I'm glad to hear from many people who use them. Luckily I still have lots of things to write about so the series will continue for quite some time. I would also like to thank my family for their support of my passion for dinosaurs, George Olshevsky, whom has been my paleontological friend and mentor, Darren Tanke who, along with George, has helped me in getting my illustrations published, the paleontologists who have helped me over the years: Jim Kirkland, and Ken Carpenter (who both helped me become a published 'paleontologist'), along with Tom Demere at the San Diego Natural History Museum, Bob McCord, Debbie Boaz and all the rest at the Mesa Southwest Museum (now called the Arizona Museum of Natural History), Dan Chure, Ralph Molnar, Peter Galton, Phil Currie, Jack Horner, John McIntosh, Spencer Lucas, Mike Brett-Surman, Don Glut, Greg Paul, Stephen Czerkas, Mark Hallett, and I apologize to all the others that I haven't named.

Preface

For as long as I can remember I've been interested in dinosaurs. No one book, toy or movie started my interest in dinosaurs but when I was a child I could find dinosaur toys or books practically wherever I went. When the Los Angeles County Library moved their books around, I still managed to find the dinosaur books, which were put up on the top shelf. I guess everything "dinosaur" had an equal interest for me. I also liked to draw dinosaurs, as well as lizards, birds, fish, crabs, octopi, etc. For the life of me, I don't know why but I didn't shade the drawings. I always just drew the outlines.

When I graduated from High School I didn't know how to continue my interest. My High School kept trying to get men into Oceanography, even though I kept saying I wanted to study Paleontology. By luck, one day in 1978 (I had graduated in 1976), my family went to the Los Angeles Natural History Museum (we had moved to the San Diego area in 1970). In the bookstore was a small book written by George Olshevsky. It had all the dinosaur names known at that time, many of which I hadn't heard of. I wrote to his Toronto address but at first did not receive a reply. However, months later George did write back. It turned out he was living in San Diego, and not far from where I lived. I had asked him about some of the dinosaurs, and specifically about *Compsognathus corallestris*. I saw an illustration of it in a book in which it had paddles for front legs. He told me he was having trouble finding the time to go to libraries to look up the new dinosaurs. I told him I could do that for both of us, and send him photocopies of the articles. After a while, not only did I send articles I found to him, but also to several paleontologists.

This started a long friendship. In 1984 George sponsored me so I could join the SVP (Society of Vertebrate Paleontology). We went to the SVP meeting held at Berkeley. It was there that George introduced me to many paleontologists and dinosaur artists that I had read about. Since then I've only missed a few SVP meetings, due to my work schedule and finances. George started to publish a newsletter for prehistoric animal enthusiasts called Archosaurian Articulations, and he let me do the illustrations for it. I had to learn pen and ink, since I had previously only worked in pencil. My father introduced me to stippling, and I taught myself that art style. Darren Tanke also asked me if I would make some illustrations for him. My biggest break in illustrating dinosaurs was creating the majority of art for the *Dinosaur Society Dinosaur Encyclopedia,* and illustrating George's articles for Gakken Mook's *Dinosaur Frontline*. Thanks to George (to whom I am eternally grateful) and all the paleontologists I've met and had the pleasure of talking about dinosaurs with, my once hobby is finally paying off, both scientifically and monetarily.

Why write a book or a series of articles on drawing dinosaurs? As I stated before, I have always been interested in dinosaurs and it wasn't until after I had visited museums to see the mounted specimens that I found out what was actually known about them. To my dismay, I found out that many of the specimens I had thought were complete, weren't, and that some had been mounted incorrectly. After doing library research I started to find out what was known about dinosaur skeletons and how the bones fit together. Another big help was being able to go to symposiums and talk to paleontologists and fellow artists. It was through these experiences that the dinosaurs, and other fossil animal life, came to life for me. This knowledge is what I want to pass on to artists and laypersons to help them to understand these wonderful animals better through their bones and articulation. There are a lot of good artists that don't know how prehistoric animals looked, anatomically speaking, and this book will help them to correct their art form. Also, there are a lot of publishers who don't use artists who know this material, and this book will hopefully help them also. This is the first in a series of books. Each book will cover 25 articles (except for the first one, which will also cover an article I did for a different publication.) In each volume I will have editorial notes that will update the subject of that article.

Please visit my two websites; http://www.dinohunter.info and http://www.paleofile.com

FOR THE DINOSAUR COLLECTOR AND ENTHUSIAST

PREHISTORIC TIMES

November-December. No, 21 $4.95

Artist Mark Hallett

Ford, T. L., 1996, How to Draw Dinosaurs. Sauropod feet: Prehistoric Times, n. 21, p. 14-15.

Chapter 1

Sauropod Feet

Before launching into the premiere installment of "How to Draw Dinosaurs", let me first explain what this column will be. I won't be telling you to "draw a large oval for the body, a small oval for the head," and so forth. What I will be telling you about is dinosaur anatomy. If you do not have the anatomy right, no matter what medium you use, the artwork will be wrong. That's something I want to correct. I'll be dwelling on matters that some may consider trivial, or unimportant, but it is the attention paid to small details that indicates that the artist has taken his or her time, and done his or her anatomical research. Some of my moments may confuse some people because many museums with mounted dinosaur skeletons have them mounted incorrectly, and it is difficult for people studying such skeletons to draw their dinosaurs right. if you have any comments, pro or con, or questions, I'll be glad to read them and, I hope, to answer them. With that said, le me now make some remarks about sauropod feet.

In all sauropods, the fore and hind feet were pentadactyl-5 toes, no fewer--, but that is about as far as the forefoot-hindfoot similarity extends. Sauropod forefeet were fully digitigrade; the animals walked on the tips of their metacarpals. Their hind feet, however, were only semi-digitigrade: they walked flat-footed on the soles of their feet, with the metatarsals only slightly elevated. Sauropod forefeet had only one claw, a large one on the first digit (or thumb), not 2-5 as some reconstructions and museum mounts show (figures 1, 2). There is footprint evidence that certain brachiosaurids(?) had lost the front claw altogether. Some have argued that no forefoot prints show the claw, but such prints have, in fact been found (figure 3). The forefoot claw was unusually large, but in brachiosaurids it was reduced (figure 6), so it is not surprising that brachiosaurid forefoot prints often show no claws.

Among the genera within any particular sauropod family, the length of the metacarpals was fairly constant, but between families their length varied considerably. They were the longest in brachiosaurids, shorter in camarasaurids and titanosaurids, and shortest in cetiosaurids euhelopodids, and diplodocids. Seen from above, the metacarpals were arranged in a hoof-shaped or horseshoe-shaped pattern, with the hollowed-out pocket behind surrounded by the metacarpals in the front and to the sides. Metacarpal III was the middle toe, with the other organized more or less symmetrically on either side of it (figures 2, 1). No pads covered the forefeet, so their prints were U-shaped hollowed out behind. These are very important points to be remembered when rendering an accurate sauropod.

Sauropod hindfeet had at least three claws, on digits I-III, the first (innermost) being the largest; they decreased in size toward the outside of the foot. I suspect, form looking at the footprint record, that sauropods may have had a claw on digit IV as well (*Dyslocosaurus polyonychius* may have had claws on all five hind-foot digits). The toes curl outward in most footprints, though they curl inward in a few (figure 3). The metatarsals formed a short semicircle, and there was a pad or cushion that served as a heel (figures 4, 5). Most of the weight was carried on their hind feet and not the forefeet.

Key points, 1), Sauropods had only one claw on the forefoot, on the "thumb"; 2) the forefoot did not have a pad, and the metacarpals were arranged in a horseshoe-shaped configuration hollowed out behind; and 3) for real accuracy, turn the hind-foot claws outward.

I would also like to thank George Olshevsky, who corrected the grammar and virtually rewrote the whole thing! (Hopefully next time I'll make it easier for him).

Tracy L. Ford PO Box 1171 Poway, Ca. 92074 dino.hunter@cox.net

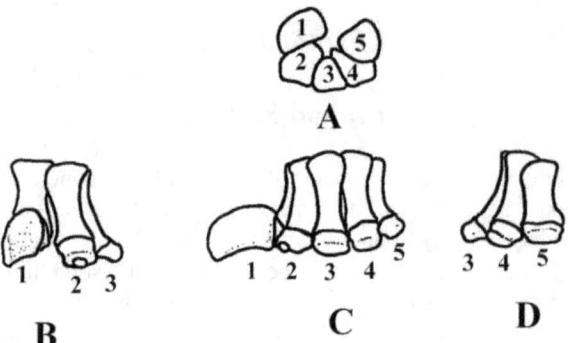

Figure 1). Left front foot of *Apatosaurus louisae*. A) Top view of metacarpals; B) Inside view; C) Front view; D) Outside view.

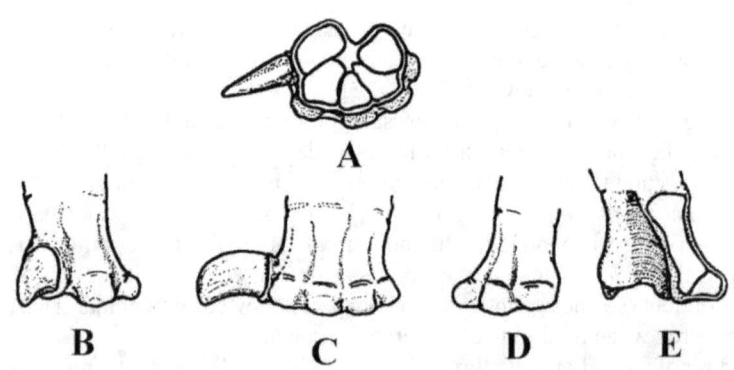

Figure 2). Left front foot of *Apatosaurus louisae*, 'fleshed out'. A) Top view of metacarpals; B) Inside view; C) Front view; D) Outside view, E) Cut away view of the third metacarpal showing the 'hollowed out pocket' that forms the 'hoof' pattern.

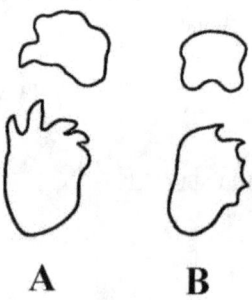

Figure 3). A) Foot prints from Galinha site, Portugal, showing the manus claw; B) *Bontopodus* form the Early Cretaceous of Texas, showing no manus claw impressions which doesn't mean there was no claw.

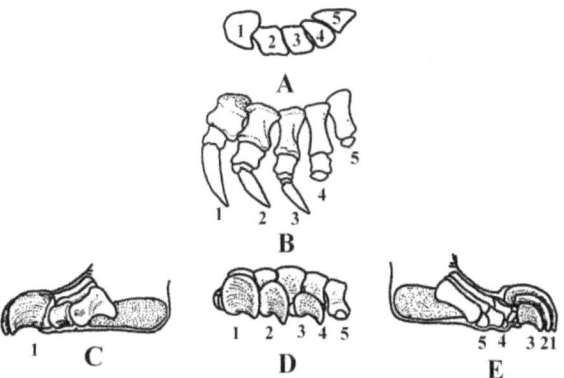

Figure 4). Left hind foot of *Apatosaurus louisae;* A) Top view of metatarsals and looking at the proximal ends of the metatarsals; B) Inside; C) Front view; D) Outside view.

Figure 5). Left hind foot of *Apatosaurus louisae;* A) Top view of metatarsals and looking at the proximal ends of the metatarsals; B) Inside; C) Front view; D) Outside view; E) Cut away view of the third metatarsal showing the 'foot' pad.

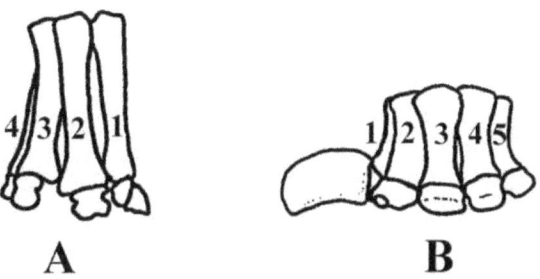

Figure 6). A) Front view of the metacarpals of *Brachiosaurus*; B) Front view of metacarpals of *Apatosaurus*.

Bibliography

Gilmore, C. W., 1936, Osteology of *Apatosaurus*, with special reference to specimens in the Carnegie Museum: Memories of the Carnegie Museum, v. 11, n. 4, p. 175-300.

Lockley, M. G., Farlow, J. O., and Meyer, C. A., 1994, *Brontopodus* and *Parabrontopodus* ichnogen. nov. and the significance of wide- and narrow-gauge sauropod trackways: In: Aspects of Sauropod Palebiology, Edited by Lockley, M. G., dos Santos, V. F., Meyer, C. A., and Hunt, A. P., Revista de Geociencias, Gaia, n. 10, p. 135-145.

McIntosh, J. S., 1990, Species determination in Sauropod dinosaurs with tentative suggestions for their classification: In: Dinosaur Systematics, Approaches and Perspectives, edited by Carpenter, K., and Currie, P. J., Cambridge university Press, p. 53-69.

McIntosh, J. S., 1990, Sauropoda: In: Dinosauria, edited by Weishampel, D. B., Dodson, P., and Osmolska, H., California University Press, p. 345-401.

Santos, V. F. dos, Lockley, M. G., Meyer, C. A., Carvalho, J., Carvalho, A. M. G. de, and Moratalla, J. J., 1994, New Sauropod tracksite from the Middle Jurassic of Portugal: In: Aspects of Sauropod Palebiology, edited by Lockley, M. G., Santos, V. F. dos, Meyer, C. A., and Hunt, A. P., Revista de Geociencias, Gaia, n. 10, p. 5-13.

Thulborn, T., 1990, Dinosaur Tracks. Chapman and Hall, 410pp.

FOR THE DINOSAUR COLLECTOR AND ENTHUSIAST

PREHISTORIC
TIMES

JAN. - FEB. No.22 $5.95

Ford, T. L., 1997, How to Draw Dinosaurs. Stegosaurus: Prehistoric Times, n. 22, p. 22-23.

Chapter 2

Stegosaurus

On occasion I come up with an idea about a particular dinosaur. After a little literature research, I might find my idea to be wrong but other times I'd find it to be worth merit. Sometimes I would be vindicated as I was regarding the spikes and plates of Stegosaurus.

I had doubted the positioning of the spikes of *Stegosaurus* for quite some time, but it wasn't until I visited the Utah Museum of Natural History in Salt Lake City that it finally came to me. I was looking at the *Stegosaurus* mount from the rear when I noticed what a large gap there was between the base of the tail spikes and caudal vertebrae (figure 1a). The mount is the typical pose with the spikes in a "V" position. Either there was cartilage or perhaps 'fat' between the base of the spikes or the spikes were mounted wrong. I thought perhaps the base of the spikes should be flush against the caudal vertebrae.

After I got home I visited my dino friend George Olshevsky and told him my new revelation (it just so happened that he was working on his Stegosaurian article for Gakken Mook's Dino Frontline). He thought it was possible but was not convinced, he needed a 'push'. That push happened just a few months later.

My 'push' came one day when a man by the name of Brian Small was out prospecting for fossils in a gully at Garden Park. With a blow of his pick into what seamed to be nothing but bare ground heard a noise that he new was bone, not rock. What he found was the second 'road kill' *Stegosaurus stenops*. This specimen is the most complete *Stegosaurus* to be found. For the first time, both the plates and spikes were found to be in place.

The spikes were discovered lying along side the caudal vertebrae (figure 1b). The first pair of spikes are the larges and angled with the tips slightly outward second pair of spikes are angled more toward the tail with the tip of the caudal vertebrae bending downward (figure 1c). An interesting note is that one of the smaller spikes was found to be pathologic, it had been broken and and slightly upward. The healed. Thanks to the work of Ken Carpenter at the Denver Museum of Natural History, the true orientation of the tail spikes are known (I've had a few personal communications with Ken Carpenter about the orientation of spikes and plates, and he's confirmed my beliefs).

The spikes aren't the only problem with the tail of *Stegosaurus*; the plates have also been in question. I won't go into the parallel, alternating or single row of plates theories (The new 'road kill' has the plates in position and they are alternating) but will discuss how the plates immobilize the tail.

In nearly all mounted stegosaurs, the tail is positioned dragging on the ground (Denver Museum of Natural History and the *Stegosaurus ungulatus* mount at the Peabody Museum are the only ones that I know of which have the tail horizontal). The caudal centra have a downward turn to them, but the last sacral vertebra is beveled upward. Both of these angled centra are what make the tail horizontal (figure 1e). Wouldn't the large plates be too heavy thus making the tail dip downward? No, it is BECAUSE of the plates that help keep the tail horizontal . Why?

I'm using the sacral and caudal vertebrae of Gilmore's reconstruction of the first 'road kill' specimens to illustrate this (figure 2). The tow largest plates, the first one over the pelvis, the second over the first part of the tail, both cover 9 vertebrae, the third largest covers 7 vertebrae. What this does is act like the ossified tendons of theropods, hadrosaurs, and ceratopians to strengthen and hold the vertebrae still. In fact, the plates are rigged and don't move, while the tendons, though ridged did allow some bending. The plates would attach deeply into the skin. A tail without the plates would have been very mobile, but the plates would lock the tail in place (figure 3 & 4). It couldn't move horizontally or vertically, since it would displace the plates out of the skin.

The tail wasn't as mobile as some moves, cartoons and drawings would make them out to be. It should be noted that *Stegosaurus stenops*'s plates, which are the largest, held the tail in the greatest restraint. The stegosaur genera with smaller plates would have had much more mobile use of the tail and be able to defend itself in a manner the public is more use to seeing in the media.

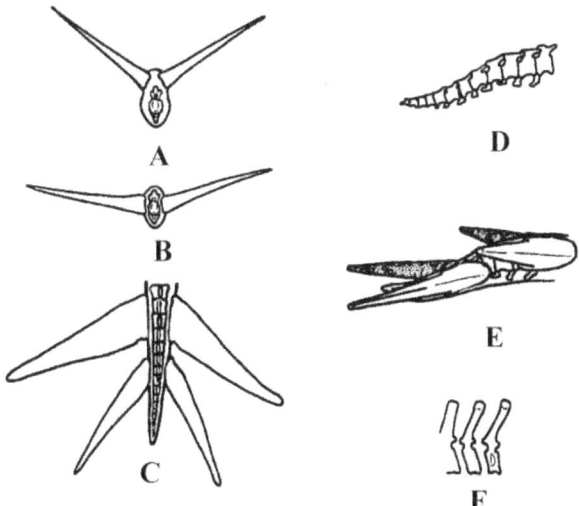

Figure 1). A) Showing the typical reconstruction of the caudal spikes from the rear; B) The corrected new interpretation: C) Looking down at the caudal spikes; D); Last caudal vertebrae showing downed bend, without the spikes; E) Last caudal vertebrae showing downed bend, with the spikes, and F) Showing the last two sacral vertebrae and the first caudal vertebrae;

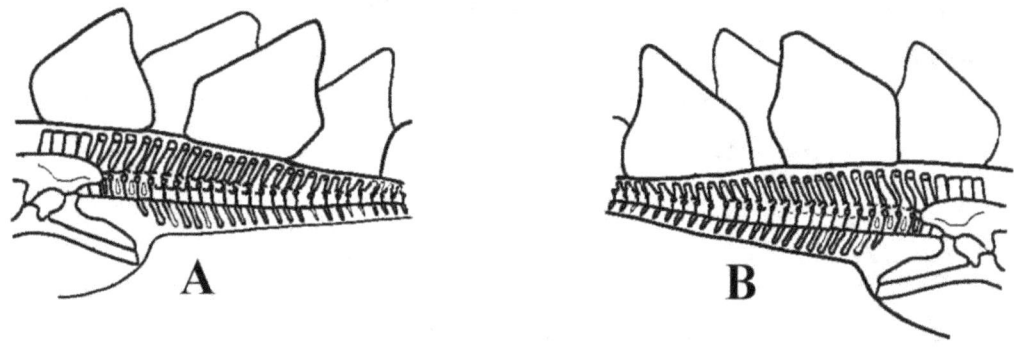

Figure 2). Right side of tail: B) Left side of tail.

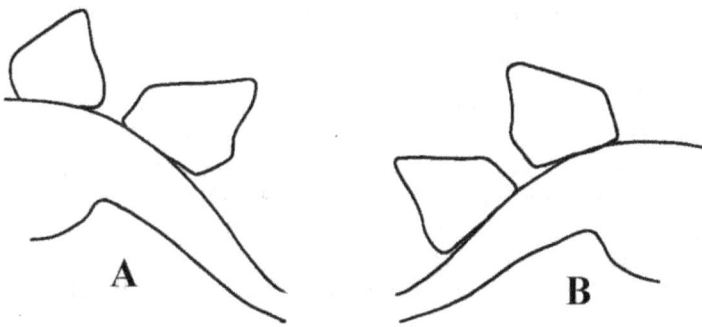

Figure 3). Right side with tail dipping down, showing the displacement of plates; B) Left side showing the same.

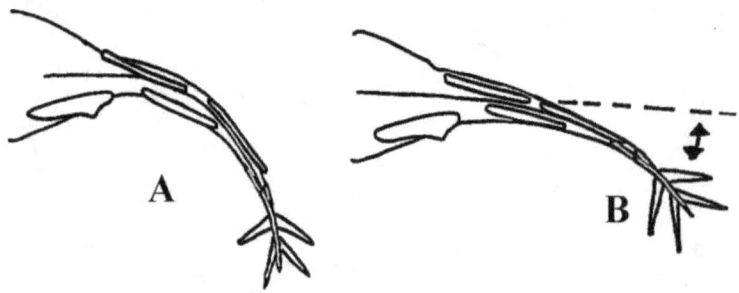

Figure 4). Top view of tail showing displacement of plates if the tail is drawn with a large arc; B) Showing the 'maximum' arc of tail.

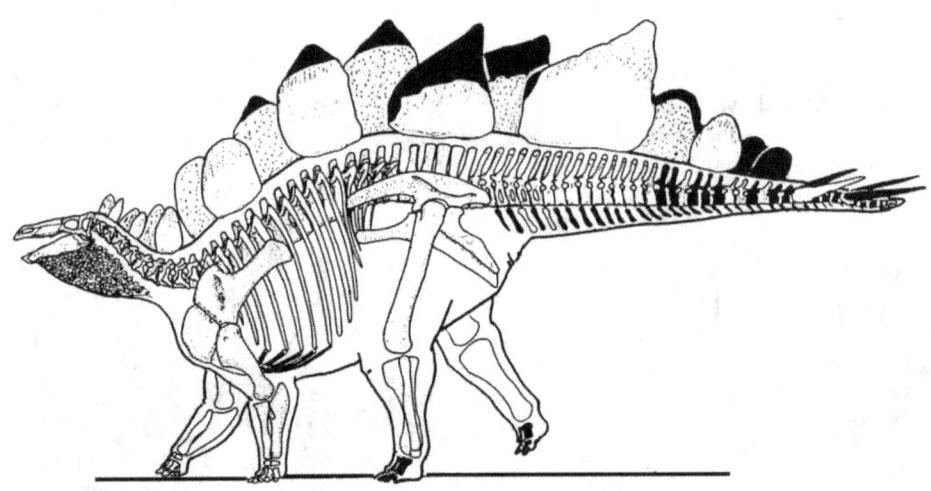

Figure 5) Skeletal restoration of *Stegosaurus stenops* (USNM 4934), new reconstruction.

Bibliography:

Anonymous, 1992, Colorado Dinosaur Discoveries: The Dinosaur Report, Fall, 1992, p. 1.

Carpenter, K., 1993, New evidence for plate arrangement in *Stegosaurus stenops* (Dinosauria): Journal of Vertebrate Paleontology, v. 13, supplement to n. 3, Abstracts of Papers, Fifty-Third Annual Meeting, Society of Vertebrate Paleontology, New Mexico Museum of Natural History and Science, Albuquerque, New Mexico, October 13-16, p. 28A.

Carpenter, K., 1996, A New Look at *Stegosaurus*: Dinosaur Discoveries, Issue 1, p. 6-7.

Gilmore, C. W., 1914, Osteology of the armored dinosauria in the U. S. National Museum with special reference to the genus *Stegosaurus*: Bulletin of the United States National Museum, v. 89, p.1-140.

Gilmore, C. W., 1915, A new restoration of *Stegosaurus*: Proceedings of the United States National Museum, v. 49, p. 355-357.

Gilmore, C. W., 1918, A newly mounted skeleton of the armored dinosaur, *Stegosaurus stenops*, in the United States National Museum: Proceedings of the United States National Museum, v. 54, p. 383-390.

FOR THE DINOSAUR COLLECTOR AND ENTHUSIAST

PREHISTORIC
TIMES MAR. -APR. No.23 $5.95

Ford, T. L., 1997, How to Draw Dinosaurs. The Dromaeosaurids: Prehistoric Times, n. 23, p. 28-29.

Chapter 3

The Dromaeosaurids

First, a commentary on the use of the term, "Raptor" for dinosaurs. There is no dinosaur or group of dinosaurs called Raptor. No dinosaur should be called, Raptor. This is a term created in a science fiction book and movies.

Unfortunately, some paleontologist seemed to have adopted this and one in particular has gone so far as to use this term for every theropod. The dinosaur family in this media is Dromaeosauridae. The term, Raptor has been incorrectly used for many other theropod dinosaurs. Raptor is a term used for modern birds of prey i.e. Eagles and Hawks, not dinosaurs.

Also as a quick grammatical note, eve the word raptor is wrong. It should be 'raptor since it is a shortening of a name like Velociraptor.

One of the things that many artist get wrong when portraying dromaeosaurs is how the hand were held when the arms were in a resting position. Unlike the majority of theropods that held their palms facing the lower arm, the palms of Dromaeosauridae (i.e. *Deinonychus*) faced each other (Figure 1). This is the same way with birds. Why is this? There is a bone in the writs dubbed the semi-lunate carpal. The semi-lunate carpal turned the writs down and toward the body, thereby folding the hand much like that in birds. The carpal elements in theropods that lacked a semi-lunate carpal folded the hand toward the arm.

Why the semi-lunate carpal? Was it a hold over from a more avian ancestry? It could have derived from a "flapping" motion that evolved into a "Flapping" grasping/strike motion. Or it could just be convergent evolution. This is a hotly debated topic and one that I will not gen into here (But just for the record, I favor the more avian ancestry).

Other families of theropods that have semi-lunate carpals are the troodontidae, oviraptoridae, and the avimimidae. (This is list is from a post from Greg Paul). The semi-lunate carpal isn't the only element that folded the arm like a bird, the humerus, scapula, coracoid etc., all helped to fold the arm.

An easy way to see these differences is to hold your arm out with your hand horizontal. Now relax your hand. You will find that your had drops with the palm toward your arm. The theropods with semi-lunate carpals were much different.

Now hold your arm and twist your pal vertical (If you're using both arms, your palms will be facing each other). Now the hard part. Turn your writs as far as you can with your fingers facing the ground, whiled still keeping your had vertical. Dromaeosaurids an the like could twist their hands much further. Kevin Padian demonstrated this beautifully at a Short Courses in Paleontology Symposium in 1989.

The type specimen had a larger "sickle" toe claw in an almost short semi circle. Newer specimens have much less of a curve which is more typical of the family dromaeosauridae (figure 2). Also, the hand claws are large as the large foot claw.

I've noticed that many artist draw the necks of dromaeosaurids short and stocky. This is wrong, the necks were slender and long (figure 3). Even though *Utharaptor* is a very large dromaeosaurid . There is no reason, in my opinion, to illustrate it with a short and stocky neck.

One last comment on the skeletal drawing of theropods and birds in general. I've noticed that sometimes the humerus is drawn upside down when the arm is folded against the body. The large 'bump' or deltopectoral crest should be facing down when the arm is folded

Gauthier and Padian (1985) have the humerus of *Deinonychus* this way. If the arm is lifted above the body, i.e. as in flapping, the crest would be facing upward. I've seen the upside down humerus on an *Ichthyornis* plate in the toothed bird monograph by Marsh, 1891. The arm, if I'm not mistaken, rotates to allow this when flapping.

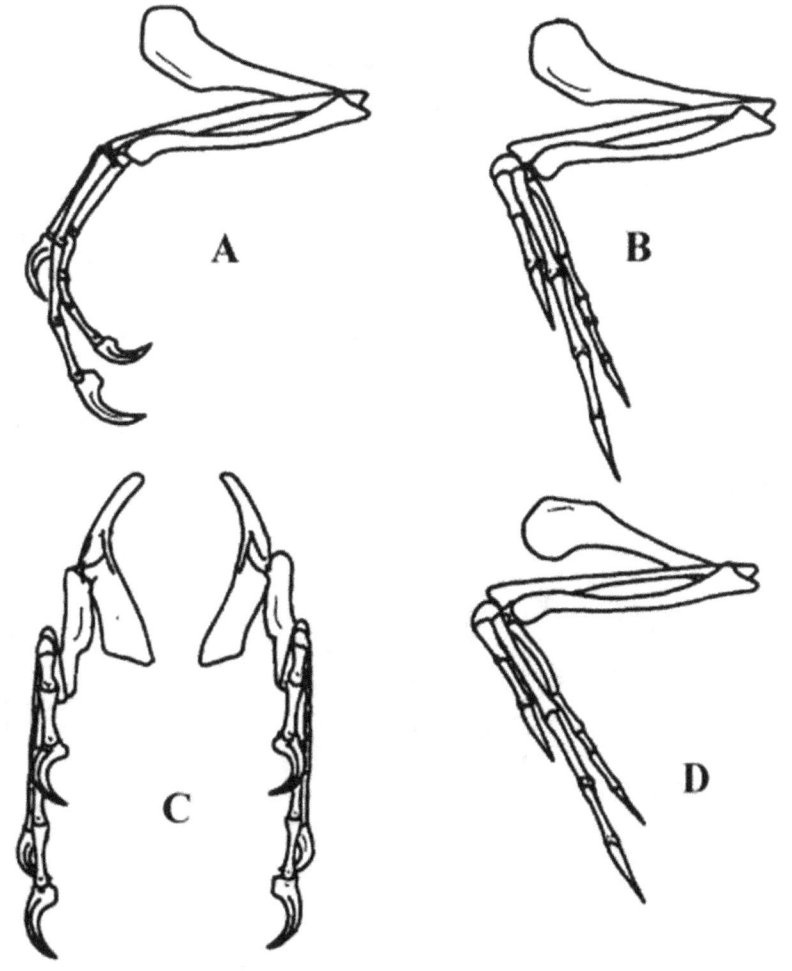

Figure 1). A) Incorrect hand; B) Correct hand; C) Wrong view of hands, and D) Incorrect humerus.

Figure 2). A) Claw from the type of *Deinonychus*, YPM 4205; B) Claw from MCZ 4371.

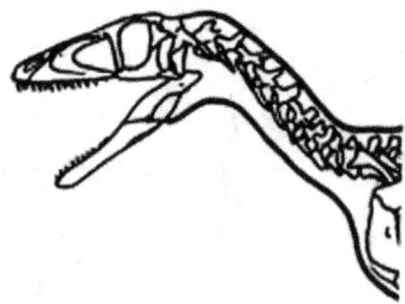

Figure 3). Neck of *Deinonychus*.

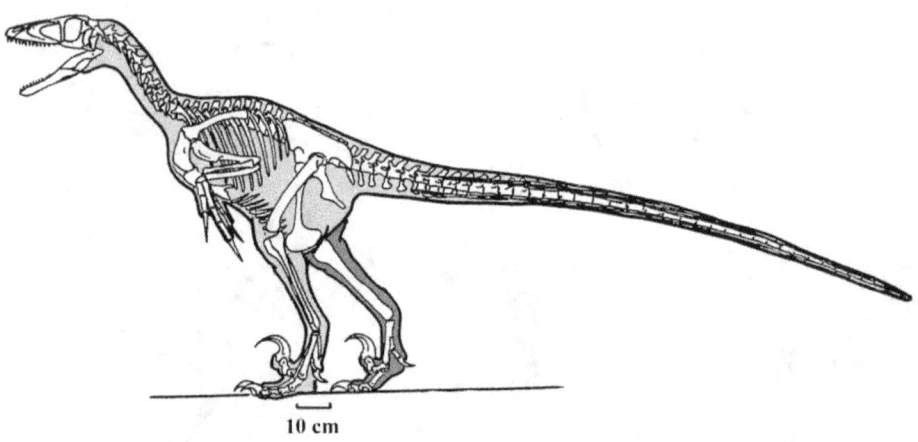

10 cm

Figure 4). Reconstruction of skeleton of *Deinonychus*.

Figure 5). Reconstruction of *Deinonychus*.
Note: This was long before feathered theropods were found from Liaoning, China, but I decide to use a corrected feathered image.

Bibliography:

Gauthier, J. A., and Padian, K., 1985, Phylogenetic, Functional, and Aerodynamic Analyses of the Origin of Birds and their Flight: In: The Beginnings of Birds. Proceedings of the International Archaeopteryx Conference Eichstatt, 1984, edited by Hecht, M. K., Ostrom, J. H., Viohl, G., and Wellnhofer, P., p. 185-197.

Gauthier, J. A., and Padian, K., 1989, The Origin of Birds and the Evolution of Flight: In: The Age of Dinosaurs. Short Courses in Paleontology, n. 2, convened by Padian, K., and Chure, D. J., series editor, Culver S. J., A Publication of the Paleontological Society, p. 121-133.

Marsh, O. C., 1881, Monograph by Professor MARSH on the Odontornithes, or Toothed Birds of North America: American Journal of Science, 3rd Series, v. 21, p. 255-276.

Ostrom, J. H., 1969, Terrible Claw: Discovery, v. 5, n. 1, p. 1-10.

Ostrom, J. H., 1969, The Supporting Chain: Discovery, v. 5, n. 1, p. 10-16.

Ostrom, J. H., 1969, Osteology of *Deinonychus antirrhopus*, an Unusual Theropod from the Lower Cretaceous of Montana: Peabody Museum of Natural History, Yale University, Bulletin 30, p. 1-165.

Ostrom, J. H., 1974, The pectoral girdle and forelimb function of *Deinonychus* (Reptilia: Saurischia): a correction: Postilia, v. 165, p. 1-11.

Padian, K., 1989, Other Mesozoic Vertebrates of the Land, Sea, and Air: In: The Age of Dinosaurs, Short Courses in Paleontology, n. 2, Convened by Kevin Padian and Daniel J. Chure. A publication of The Paleontological Society, p. 146-163..

FOR THE DINOSAUR COLLECTOR AND ENTHUSIAST

PREHISTORIC

TIMES   MAY- JUNE   No. 24 $5.95

Ford, T. L., 1997, How to Draw Dinosaurs. *Iguanodon*: Prehistoric Times, n. 24, p. 30.

Chapter 4

IGUANODON

Iguanodon and *Megalosaurus* were the first dinosaurs to be described (in 1822). *Iguanodon* is one of the most famous dinosaurs and is known from numerous areas, Europe, Mongolia, North America, North Africa etc. The problem with drawing *Iguanodon* is choosing which *Iguanodon* to draw.

Iguanodon bernissartensis and *Iguanodon atherfieldensis* are the best known species. Both species have been found in quarries with each other, but one is always more prominent than the other. Some may think that this is sexual dimorphism but since there is never a good number of each found together, this is probably not the case. (**Editors Note: Paul (2006) renamed *Iguanodon atherfieldensis* *Mantellisaurus* *atherfieldensis* (HOOLEY, 1925) PAUL, 2006, = *Iguanodon? atherfieldensis* HOOLEY, 1925. For this book I be using the old name**)

Iguanodon bernissartensis is a robust animal (figure 1a), with short metacarpals and a large thumb spike (figure 2a). From the length between the front and hind legs, I would be inclined to make *Iguanodon bernissartensis* a quadrapedal animal. The skull was also more robust (figure 3a). The dorsal vertebrae was short (figure 4a).

Iguanodon atherfieldensis was a gracile animal (figure 1b), with longer metacarpals and a small thumb spike (figure 2b), and was probably a bipedal animal judging by the length differences between the front and hind legs. The skull was also more gracile (Figure 3b). The dorsal vertebrae was taller (figure 4b).

There are only two claws on the hands (not including the thumb spike) of *Iguanodon*. The second and third metacarpals have the claws, that is if we assume that the thumb spike is the first metacarpal.

There are other species of *Iguanodon*, but these are the ones with the most material. Something to take into account is the belly gastralia (stomach ribs) found with any Ornithischian. This caused the belly to hang lower due to the large digestive track, i.e. stomach and intestines. You can either draw them with a 'slim' belly or extend it some.

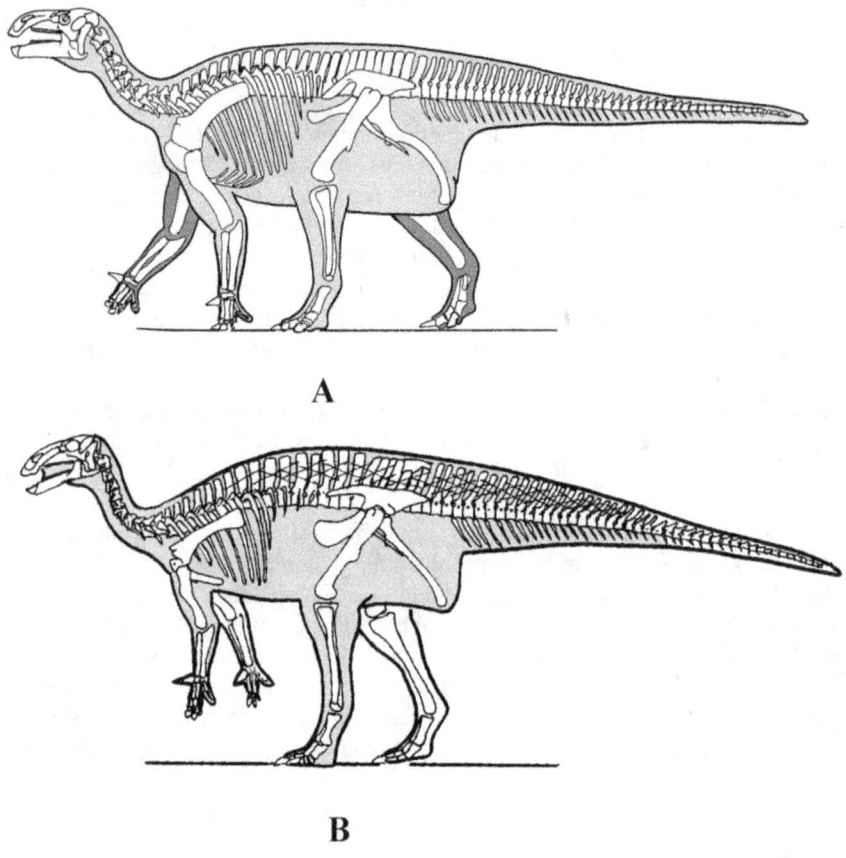

Figure 1), A) *Iguanodon bernissartensis* skeletal and life restoration; B) *Iguanodon atherfieldensis* skeletal and life restoration. Not drawn to the same scale.

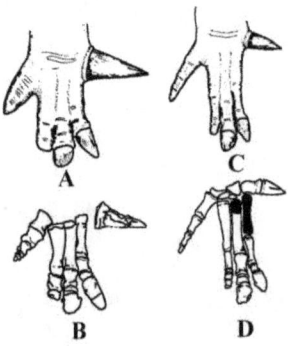

Figure 2), A) *Iguanodon bernissartensis* manus in skeletal and reconstruction; B) *Iguanodon atherfieldensis* manus in skeletal and reconstruction. Not drawn to the same scale.

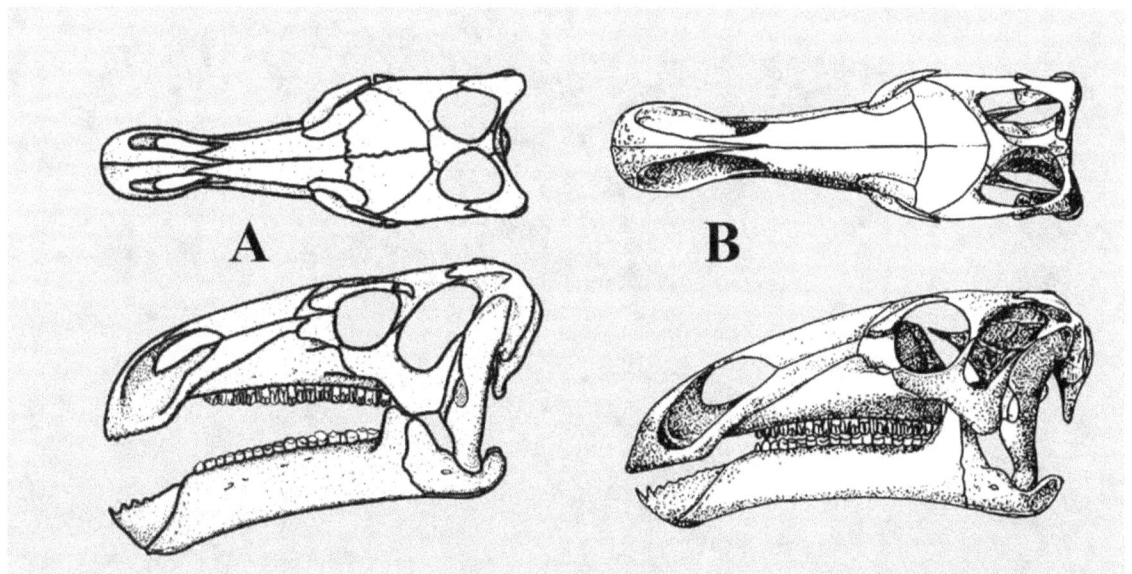

Figure 3), A) *Iguanodon bernissartensis* skull, top and side; B) *Iguanodon atherfieldensis* skull, top and side. Not drawn to the same scale.

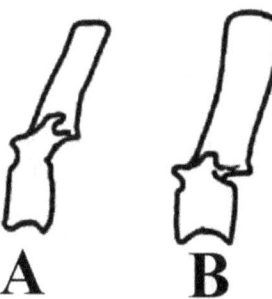

Figure 4), A) *Iguanodon bernissartensis* dorsal vertebrae; B) *Iguanodon atherfieldensis* dorsal vertebrae. Not drawn to the same scale.

Figure 5), Life restorations; A) *Iguanodon bernissartensis* skeletal and life restoration; B) *Iguanodon atherfieldensis* skeletal and life restoration. Not drawn to the same scale.

Bibliography

Norman, D. B., 1980, On the ornithischian dinosaur *Iguanodon bernissartensis* of Bernissart (Belgium): Institut Royal des Sciences Naturelles de Belgique, Memore, n. 178, p. 1-103.

Norman, D. B., 1986, On the anatomy of *Iguanodon atherfieldensis* (Ornithischia: Ornithopoda): Bulletin del l'Instut Royal Des Sciences Naturelles de Belgique, Sciences de la Terre, v. 56, p. 281-372.

Paul, G. S., 1987, The Science and art of restoring the life appearance of Dinosaurs and their relatives: In: Dinosaurs Past and Present, v. 2, p. 4-49.

FOR THE DINOSAUR COLLECTOR AND ENTHUSIAST

PREHISTORIC TIMES

AUG - SEPT NO. 25 $5.95

Artist Douglas Henderson

Featuring Bob Bakker, Gregory Paul, Peter Larson and The Lost World

0 74470 92325 1   08>

Ford, T. L., 1997, How to Draw Dinosaurs. Give Theropods no Lip!: Prehistoric Times, n. 25, p. 49-50.

Chapter 5

Give Theropods no Lip!

I've seen Robert Bakker talk on a lot of documentaries, and in several of them he said 'think lips' for theropods. He clams theropods have lizard-like lips. Many artists have depicted theropods this way, before and after what Bakker has said. Well, I have 'thought' about this, and in the last month, I've thought about it a lot, and what I've determined is that theropods couldn't have had lips.

Nanotyrannus is a good dinosaur to use. It has its mouth closed and shows it's maxillary teeth (though the majority of these teeth are fake) (figure 1).

The dentary of dinosaurs fits inside the upper jaws (figure 4). In theropods, the surangular fits inside the supratemporal fenestra (the large 'hole' that is the inside of the quadratojugal, and jugal), the dentary then fits inside the maxilla and the tip of the dentary fits inside the premaxilla. The premaxilla in tyrannosaurids is small, but in all other theropods the premaxilla is fairly large. The maxilla of most theropods lies above the jugal, while in *Tyrannosaurus* it lies below the maxilla. The reason for this is that the snout is downturned, the frontal-nasal area is thin, and the head is held at a 45° angle because of the orientation of the occipital condyle (where the neck and head connect). This means that the head of *Tyrannosaurus* is also held at a 45° angle when relaxed, and it is at the thinnest part of the upper portion of the skull that *Tyrannosaurus* mainly sees with its stereoscopic vision. For instance if you draw a *Tyrannosaurus* with its head at a 45° angle, then draw a level line (to the bottom or top edge of the paper) you'll see just where the narrowest part of the skull is, and that is where the *Tyrannosaurus* stereo vision is. (George Olshevsky told me this. Give credit where credit is due I always say). In *Nanotyrannus* the maxilla lies just above the maxilla.

Even though the jaws of *Nanotyrannus* are closed and can't be opened this doesn't mean that the height of the palate and length of the dentary teeth can't be figured out (figure 2). This is an important element in the discussion of lips. The dentaries upper edge lines up (inside) just a little above the maxilla margin. It is either the height of the palate that dictates the length of the dentary teeth or the dentray teeth that dictates the height of the palate. The palate itself lies just below the nasal and just below the antorbital fenestra. The palate in *Tyrannosaurus*, *Tarbosaurus* and other tyrannosaurids has little 'pockets' for the larger dentary teeth to fit into (no telling if this is the case of *Nanotyrannus*). All dinosaurs have an overhanging upper jaw, or an overbite, because the lower jaw does fit inside the upper jaw. This overhang is due to the width of the teeth, plus the width of the adjoining bone, and the ventral edge of the jugal, and quadratojugal.

The dentary doesn't even lie within the premaxilla. If a theropod or any other dinosaur for that matter, has a dentray that lines up with the premaxilla, then that is either drawn, reconstructed wrong or has the wrong dentary. For example, the premaxillary teeth follow the tip of the dentary, if the teeth are at a strong angle, similar to that of USNM 4734 (*Allosaurus fragilis*) (figure 3). The premaxillary teeth are at a sharp angle, and the dentary lies up with the premaxilla. This is wrong. There is a lower jaw missing the tip of the dentary (type of *Labrosaurus ferox*, USNM 22315) found a meter away from the skull, that may belong to the skull and not the dentary that is with the skull. This is what George Olshevsky has told me and I agree with him. Since the tip is broken, or maybe more properly has been bitten off, then the tip of the dentary couldn't guide the premaxillary teeth straight so that is why the teeth are at such a step angle.

There is a line of little fossa along the margin of the premaxilla, maxilla and dentary. These fossa are what Bakker calls 'lip holes' (figure 6). These 'lip holes' haven't anything to do with lips. What they are for are blood vessels and nerve endings (which Bakker does agree with). What these nerve endings and blood vessels do is to help in the teeth replacement and has little to do with gums. The teeth I nearly all dinosaurs (except for *Heterodontosaurus*) are constantly being replaced. They are not in the jaws long enough to get tooth decay. Crocodilians replace their teeth about every 8 months. In dinosaurs that time would probably have been longer. I won't go into how the teeth are replaced, that is a subject for a different forum.

Lizard jaws also fit inside each other. Lizards don't have lips per say, what they have is an end of skin along the mouth edge that has large scales forming the edge. When the mouth is open this edge, depending on the lizard genus, hangs just below the teeth or above the teeth. When the mouth is closed the edges compress each other forming a tight fit.

Now to the theropods (figure 5). If theropods did have lips, then the upper lips would have lined up at or just below the premaxilla, maxilla, jugal, and quadratojugal. The outside of the lower jaws' lip would have had to have been the width of the maxillary teeth and the adjoining bone, forming a grove for the upper teeth to fit into. Some theropods' the upper teeth even extended past the lower edge of the dentary, and if they had a lower lip then they would have bitten through the lower jaws. The front of the lower lips would have had to extend well beyond the tip of the dentary in an even bigger grove.

What did the edge of the upper and lower jaws look like? Since crocodilians are the closest archosaur that has teeth (birds being the closest) to that of dinosaurs, I think that this is what theropods and for that matter, prosauropods' and sauropods' jaws look like.

Do I think *Nanotyrannus* is a valid genus? Yes. After talking to Neal Larson (whom I saw at the Mad Model Party in Pasadena in May), and looking at *Tyrannosaurus's* jaws for the more scientific article about theropods lips, *Nanotyrannus* has many things that are different than *Tyrannosaurus*. *Tyrannosaurus* has a maxilla that dips below the jugal. Why? The muzzle actually dips down so that the animal can use its stereo vision, the occipital condyle is a right angle, the lachrymal is straighter, etc. In *Nanotyrannus*, the maxilla is at the same level as the jugal, or just above it, the muzzle doesn't tip down, the occipital is not at a 45 degree angle, etc.

One more thing that Neal told me, which I'm very grateful for, and that is just where the ear goes. I've wondered that for some time. Neal told me that the ear lies along the edge of the quadrate. The lower end of the quadrate bows inward a little and that is where the ear is in modern reptile, birds, etc. Now we all know. But what did the ear look like? Well, that I can't tell. I don't know if it was covered, had a shallow ear canal, or a large one. But the ear didn't look like mammals.

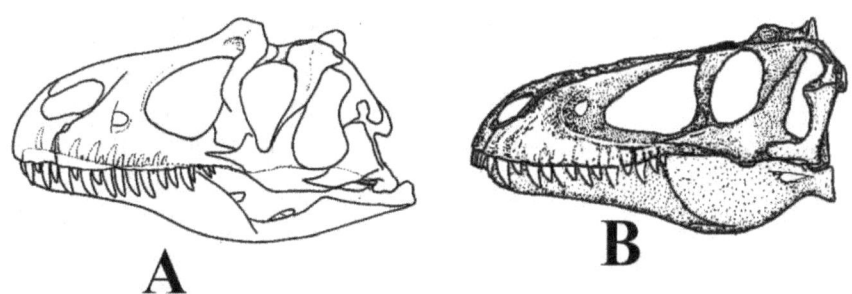

Figure 1), Side view of the skull with jaw closed; A) *Allosaurus* AMNH, B) *Nanotyrannus* (after Bakker, et al., 1988).

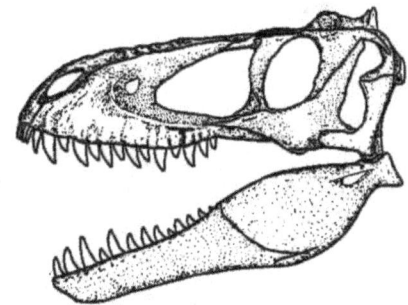

Figure 2), Side view of the skull with jaw open (modified after Bakker, et al., 1988).

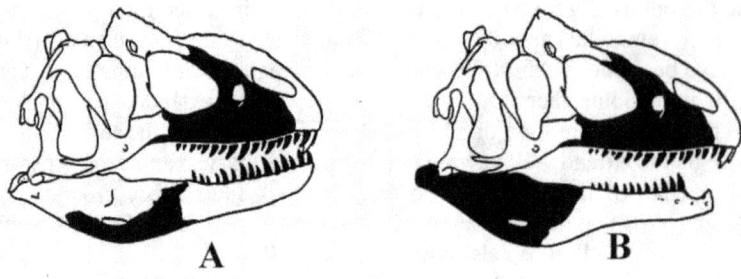

Figure 3), *Allosaurus* USNM 4734 (*Allosaurus fragilis*) side view with wrong lower jaws, b) with right lower jaw?

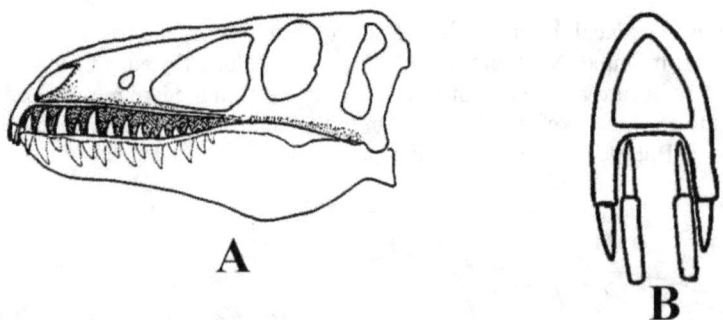

Figure 4), Cut away view of the skull showing the height of the palate and how the teeth fit into the skull; A) front view, and B) side view.

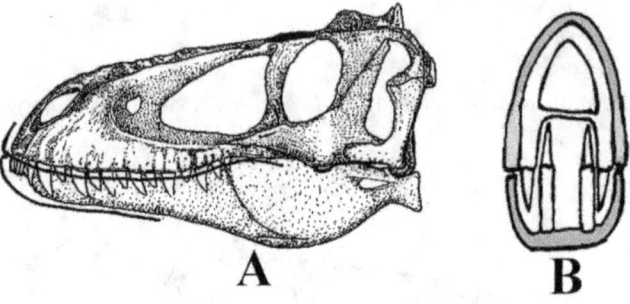

Figure 5), A) Side view of the skull with lips; and B) cut a-way of mid section, with lips added.

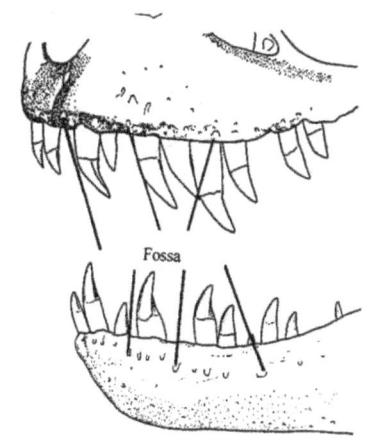

Figure 6), The line of fossa (or Bakker's lip holes) on the premaxilla, maxilla and dentary.

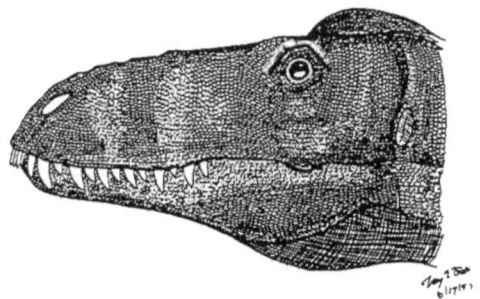

Figure 7), Life restoration of the skull of *Nannotyrannus*.

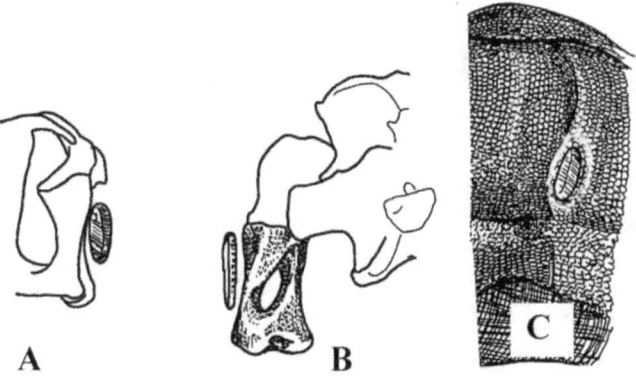

Figure 8), Position of the ear; a) back end of the skull with the ear in place, and B) caudal region of the skull showing quadrate (stippled bone), and the ear displaced; and C) showing the position of the ear.

Bibliography

Ford, T. L., 1997, Did Theropods have Lizard Lips?: Southwest Paleontological Symposium – Proceedings, 1997, p. 65-78.

Frazzetta, T. H., 1983, Adaptation and function of cranial kinesis in reptiles: a time- motion analysis of feeding in alligator lizards: In: Advances in Herpetology and Evolutionary Biology, Essays in Honor of Ernest E. Williams, edited by Rhodin, A. G. J., and Miyata, K., p. 222-244.

Gilmore, C. W., 1920, Osteology of the Carnivorous Dinosauria in the United States National Museum, with special reference to the genera *Antrodemus* (*Allosaurus*) and *Ceratosaurus*: Bulletin of the United States National Museum, n. 110, p. 1-159.

King, G., 1996, Reptiles and Herbivory: Chapman and Hall, 160pp.

Madsen, J. H. jr., 1976, *Allosaurus fragilis* a revised osteology: Utah Geological and Mineral Survey, Bulletin, n. 109, p. 1-163.

Molnar, R. E., 1973, The Cranial Morphology and mechanics of *Tyrannosaurus rex* (Reptilia; Saurischia). Master Thesis: University Microfilms International, 73-18,639, 451pp.

Witmer, L. M., 1997, The Evolution of the Antorbital Cavity of Archosaurs: A Study in Soft-Tissue Reconstruction in the Fossil Record with an Analysis of the Function of Pneumaticity: Journal of Vertebrate Paleontology, v. 17, supplement to n. 1, memoir 3, p. 1-73.

FOR THE DINOSAUR COLLECTOR AND ENTHUSIAST

PREHISTORIC TIMES

Oct. – Nov. 1997 • No. 26 • $5.95

T. REX
vs.
GIGANOTOSAURUS

J.H. MILLER
COLLECTIBLES

ARTIST JOHN SIBBICK

SIBBICK 95

Ford, T. L., 1997, How to Draw Dinosaurs. Sclerotic rings? The eyes have it: Prehistoric Times, n. 26, p. 11.

Chapter 6

Sclerotic rings? The eyes have it

This is a good opportunity to talk about eyes. *Troodon* has very large eyes. Just how do dinosaur eye look? Or more importantly to this article, what about the sclerotic rings? No troodont has been found with sclerotic rings. This is a short list of dinosaurs that have been found with sclerotic rings. *Plateosaurus, Diplodocus, Nemegtosaurus, Dromaeosaurus, Dromiceiomimus, Hypsilophodon, Parksosaurus, Lambeosaurus,* Hadrosaurs, and those pesky avian theropods, Birds. For this article the differences of the sclerotic rings is not important. I don't know of all dinosaurs had sclerotic rings since they do not preserve very well.

Where does the sclerotic ring fit? On the outside of the eye or inside? Would you be able to see it? The sclerotic ring fits just inside the surface of the eye, so you wouldn't be able to see the ring. The ring supports the shape of the globe, and the cornea bulges out around the sclerotic ring. If you know how large the sclerotic ring is, and more importantly to an artist, how big the inside of the ring is, then you can accurately depict the size of the eye. It is the inside area that will have exposed eye since the cornea area, including the iris and pupil is all that would be seen. Even though the eyes are large, the part that is seen would look small, i.e. in *Lambeosaurus*. In a paper by Underwood, (1970) he cautions against just using the size of the orbit for eye size. The eye may not encompass the total orbit, and it is only the size of the sclerotic ring that can accurately determine the size of the orbit. A large sclerotic ring suggests a nocturnal habit. *Troodon* has large eyes, possibly as large as the orbit, and is thought to have had a nocturnal life, but *Dromiceiomimus*, which does have a sclerotic rings, have very large eyes, and they most likely had a nocturnal habit.

There may have also been a thin membrane that covered the eye, similar to the nictitating membrane in birds. How exactly to draw a pupil is up to the artist. Do you draw a small pupil? A large one? One like a bird, crocodile, snake? Me, I prefer a bird.

Not to at least have something to say on *Troodon*. *Troodon* has very large forward facing eyes, and a wider posterior end than large, long narrow snout.

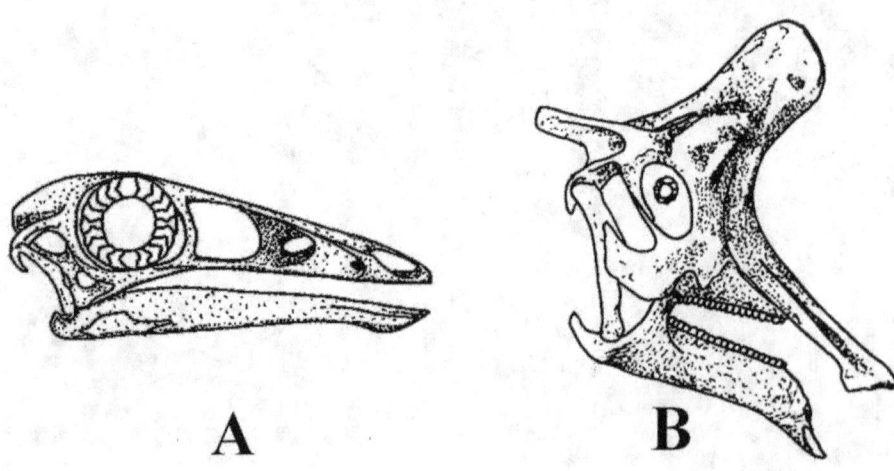

Figure 1), Skulls of A) *Dromiceiomimus samueli* ROM 840; B) *Lambeosaurus lambei*, ROM 1218, showing the different sizes of sclerotic rings.

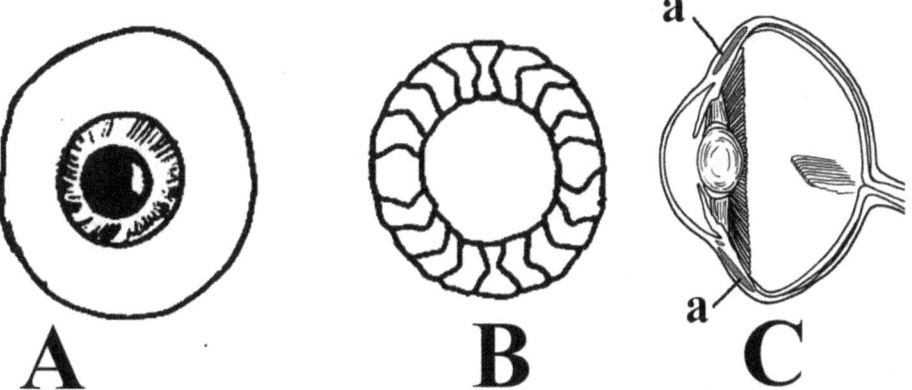

Figure 2), Hypothetical eye of *Dromiceiomimus*, A-B); A) Reconstruction, B) Sclerotic ring, and C) Cut away view of the eye, a showing the sclerotic ring in the eye is located. (Highly modified from a Roc Dove's eye, Proctor & Lynch, 1993).

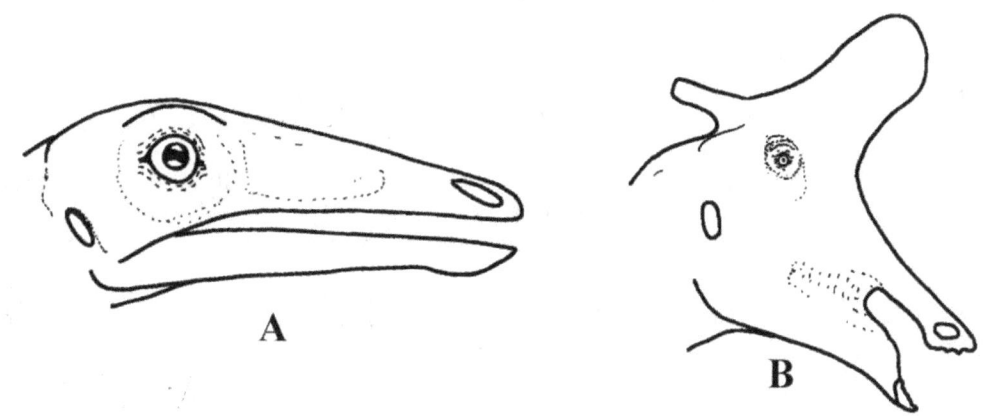

Figure 3), Reconstruction of the skull, and life restoration showing the eyes in A) *Dromiceiomimus* (the eye large in *Dromiceiomimus* see Figure 1 a for reference), and B, C) *Lambeosaurus*. Notice how small the eye looks in *Lambeosaruus*.

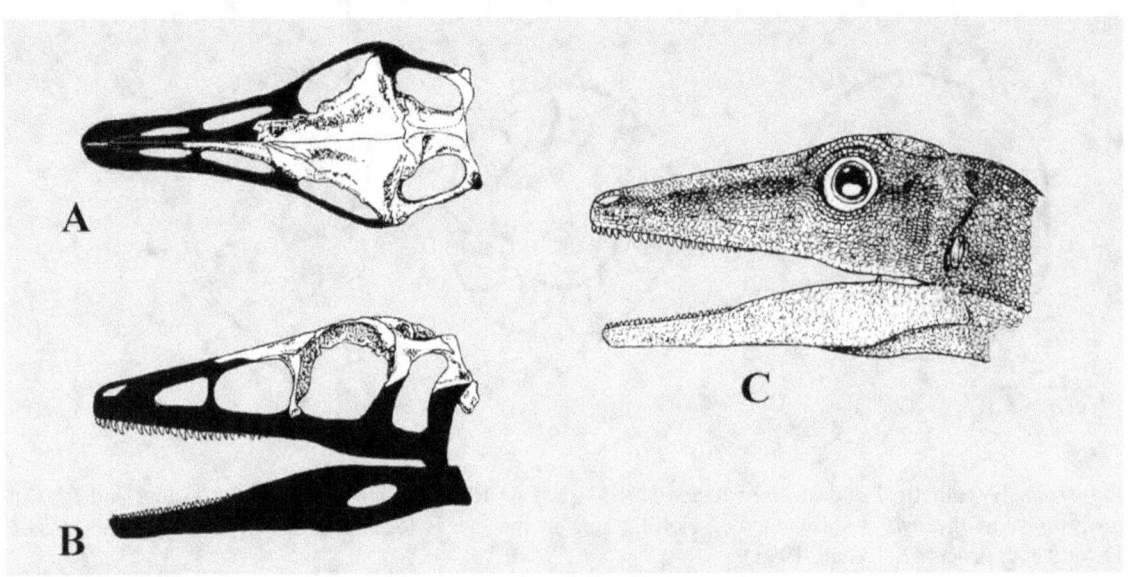

Figure 4), Skull of *Troodon*; A) Dorsal, B) Lateral, and C) Life restoration.

Bibliography

Proctor, N. S., and Lynch, P. J., 1993, Manual of ornithology, avian structure & function. Yale University Press, 340pp.

Russell, D. A., 1969, A new specimen of *Stenonychosaurus* from the Oldman Formation (Cretaceous) of Alberta: Canadian Journal of Earth Sciences, v. 6, p. 595-612.

Russell, D. A., 1972, Ostrich Dinosaurs from the Late Cretaceous of Western Canada: Canadian Journal of Earth Sciences, v. 9, p. 375-402.

Russell, D. A., and Seguin, R., 1982, Reconstruction's of the small Cretaceous Theropoda *Stenonychosaurus inequalis* and a Hypothetical Dinosauroid: Syllogeus, n. 37, p. 1-43.

Russell, L. S., 1940, The sclerotic ring in the hadrosauridae: Palaeontological Contributions of the Royal Ontario Museum, n. 2, p. 1-7.

Underwood, G., 1970, Chapter 1, The Eye. In: Biology of the Reptilia, edited by Gans, C., v. 2, Morphology B, p. 1-97.

For The Dinosaur Collector and Enthusiast

PREHISTORIC TIMES

Jun/Jul 1998 No. 30

5TH ANNIVERSARY ISSUE

CELEBRATING FIVE YEARS OF PREHISTORY

U.S. $5.95 • Canada $6.95

0 74470 92325 1

06>

Rigby's T-rex
The largest meat-eater in the world

DINOFEST 1998

Ford, T. L., 1997-1998, How to Draw Dinosaurs. Sauropods, sticking their necks out: Prehistoric Times, n. 27, p. 34-36.

Chapter 7

SAUROPODS, STICKING THEIR NECKS OUT

(Editor's note...I was told by my good friend Scott Hartman, that the necks of sauropods don't bend down as I wrote in this article. He told me, from his first hand observations of the skeletons, the illustrations I used from the monographs were wrong. I've been waiting for someone to publish on the necks of sauropods, because I want to correct this misidentification on my part. So, disregard the section on how I have the neck dipping downward, but the rest is good).

Sauropods are the most recognizable of the dinosaurs. They had a wide range of sizes, from Grey Hound size (A new specimen was mentioned at the SVP by Maxwell, Hallas, and Horner, 1997), to more than 130 feet long. The most striking feature of sauropods are their long necks. Naturally the assumption is that they ate from the treetops with their outstretched neck. But this assumption is being challenged by Stevens & Parrish. Since sauropods are so large, articulating an adult sauropod neck is both time consuming and costly; not to mention back-breaking.

What Stevens and Parrish (They are among the few who are pioneering what is called 'cyber paleontology') are doing is taking measurements of the individual cervical (neck) vertebrae and then plotting them into a computer. The idea is to try to determine just how the necks of sauropods were held and how far they could flex, with controversial results. I will be integrating their work into this article.

Sauropods necks vary in length, width and height. The cervical vertebrae has a 'ball' at the lower anterior (front) end. This 'ball' fits into a 'cup' on the posterior (back) end of the preceding vertebra. There are two prongs (prezygopophysis) on the upper anterior end that fit into the upper posterior end (postzygaphopysis) (figure 1).

This determines how far the vertebrae could move, up, down, and sideways. Also the neural spine is either single or split in two (bifurcated). There are two cervical ribs that extend below the vertebrae and extend to the end of the originating cervical, or in some cases either 1 or 2 vertebrae afterwards. These ribs are for muscle attachment and strengthen the neck. This is not to say that the neck is locked straight, into a solid straight line. The ribs could bend, possibly similar to that, but not as extreme as, the wishbone in a bird.

How long the neck is determined by how many, and how long the cervical vertebrae are. I will break down each family using the following terms for each group. The terms are; Length = short, medium, or long; width will be; slender, or wide; and height will be determined; as high, small and deep. Deep being how low the cervical ribs are. Also, whether or not the neural spine is single or bifurcated. Sauropod necks not only differ from family to family (figure 2), but from genus to genus in those families. Diplodocids neck lengths vary greatly. *Barosaurus* has the longest neck, about 1/3 longer than *Diplodocus* (and consequently, the tail of *Barosaurus* is 1/3 shorter than Diplodocus), *Diplodocus* has a medium sized neck, with *Apatosaurus lousiae* being short and *Apatosaurus ajax* the shortest (figure 3).

Cetiosaurids have 13-14 cervical vertebrae. Single neural spine, vertebrae short, slender and small.

Titanosaurids: No complete neck of a titanosaur has been described yet. New specimens have complete necks, but I don't know how many they had. I do know that they had a single neural spine, vertebrae, short, slender and small.

Brachiosaruids: No complete neck ahs been found, it is estimated to have 13 cervicals. Single neural spine, vertebrae long, slender and small.

Camarasaurids have 12 cervicals. Bifurcated neural spines, vertebrae short, wide and small.

Euhelopids (Mamenchisaurids) have 19 cervicals. Single neural spines, vertebrae long, slender and small.

Diplodocids have 15 cervicals. Bifurcated neural spines, vertebrae range in long, slender, and short.

Dicraeosaurids have 12 cervicals. Bifurcated neural spines, short, slender and tall.

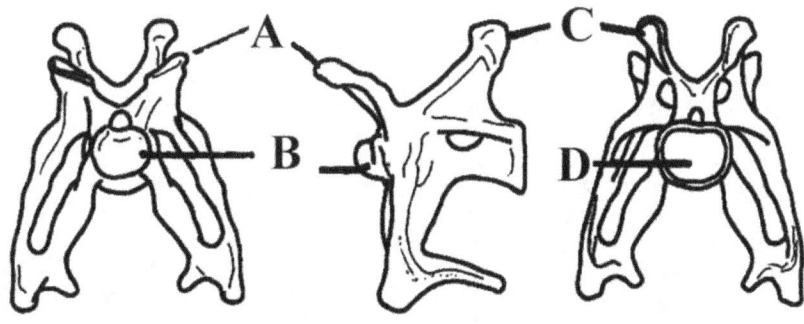

Figure 1), Cervical vertebrae (8th) of *Apatosaurus excelsus*, in front, side and back view showing A) Prezygopophysis; B) "Ball", C) "Cup", and D) Postzygopophysis.

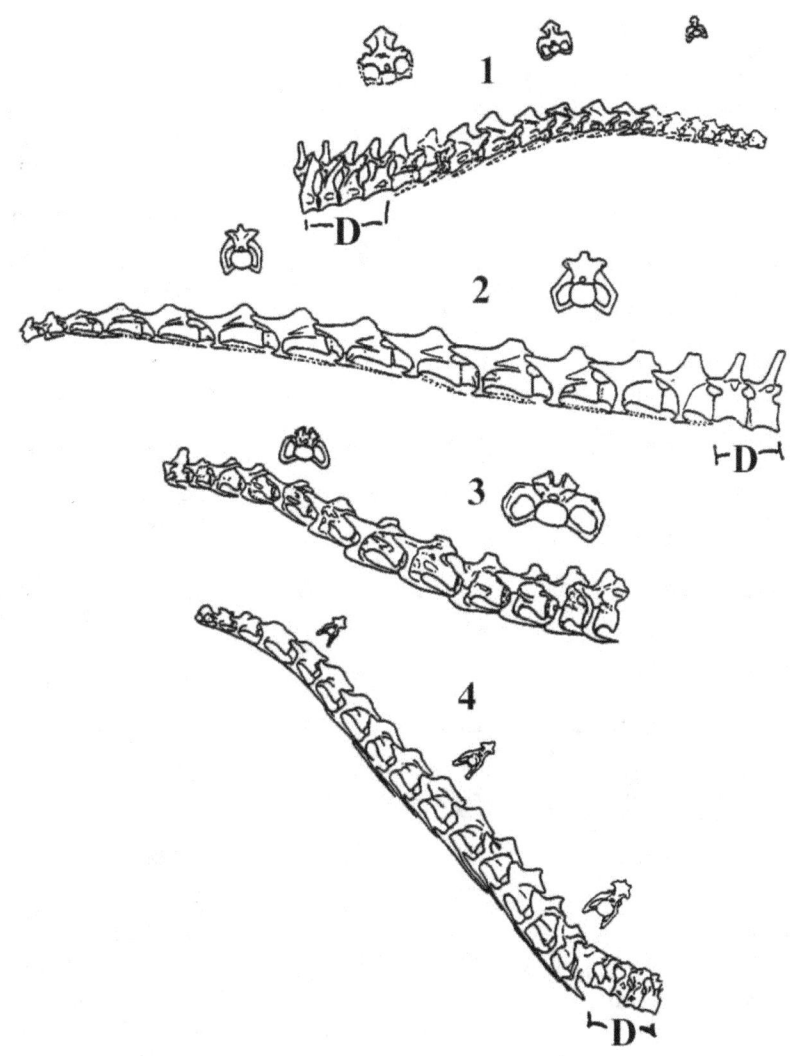

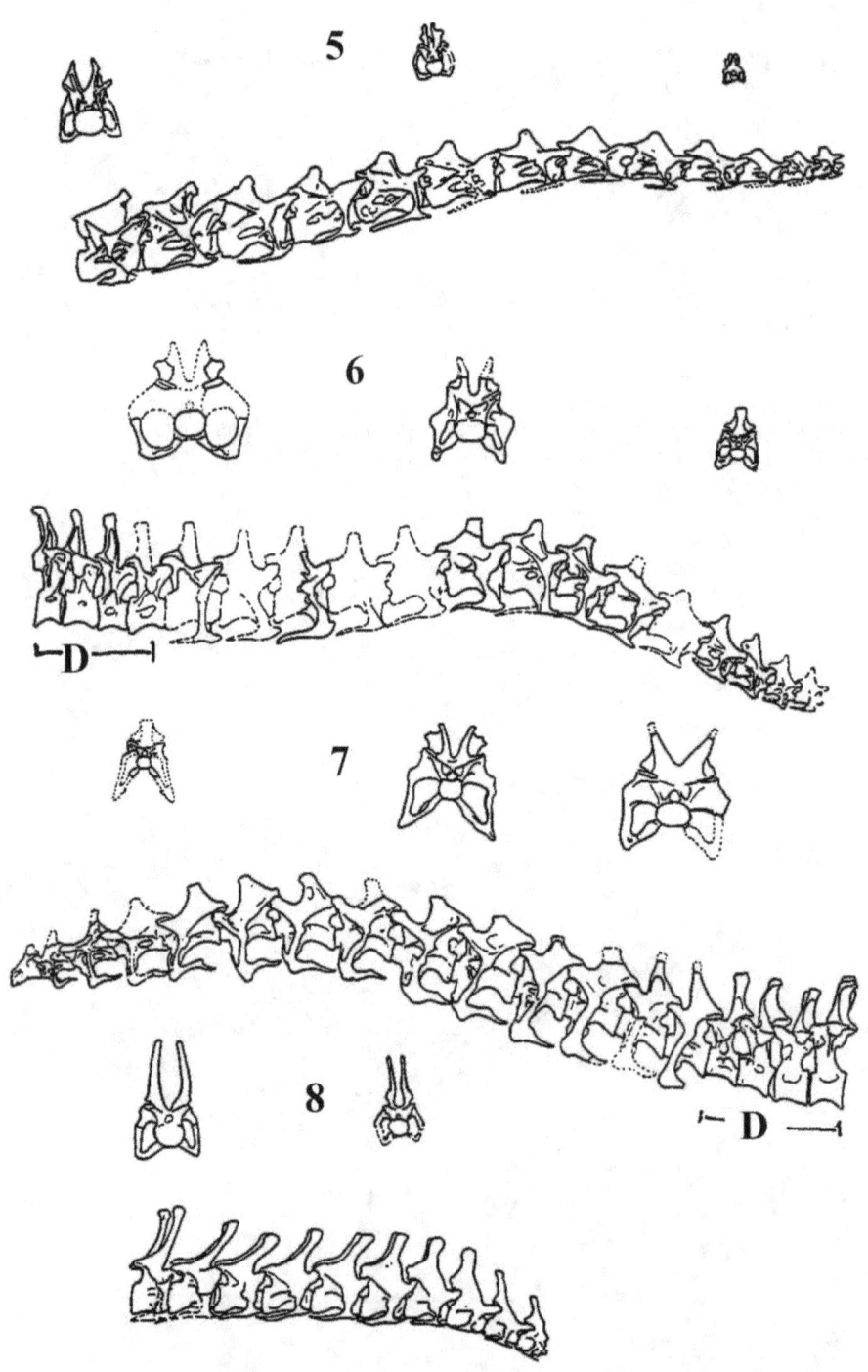

Figure 2), The necks of sauropods to how the back was held, with the front view of cervicals 5, 10 and when it is known, 15: 1) *Haplocanthosaurus priscus* with dorsals; 2) *Giraffatitan brancai* (I am modifying *Brachiosaurus* to *Giraffatitan* via Greg Paul, via George Olshevsky, 1992), with dorsals 1-2); 3) *Camarasaurus supremus*; 4) *Euhelopus zadanskyi* with dorsals 1-4; 5) *Diplodocus carnegii*; 6) *Apatosaurus excelsus* with dorsals 1-4; 7) *Apatosaurus lousiae* with dorsals 1-4); 8) *Dicraeosaurus hansemanni*.

After looking at the height, width and length, now its time to look at how they held their necks. Did sauropods hold their necks up, liked that of a giraffe? S-shaped vertical neck? Straight out? Or something in between? There are several anatomical parts of the skeleton that need to be looked into. First, are the hind legs longer or shorter than the front? In diplodocids and dicraeosaurids, the front legs are shorter than the hind. This brings the front end of the body down, with the neck. Or are the legs the same length? In camarasaurids and brachiosaurids especially, both sets of legs are the same length, this brings the neck above the height of the hind legs.

The dorsal (back) vertebrae also needs to be taken in to consideration. For the most part, the back is held in a straight ling. In *Dicraeosaurus* the back bends forward, due to the length of the legs. For the most part, the first dorsals' 1 and 2, are beveled upward, this raising the neck. In euhelopids, the neck is held very high.

Stevens & Parrish are discovering that the traditional view of how sauropods held their neck is incorrect. For their example, their research has shown that the neck of diplodocids was held, not curved upward, but straight out.

When I started this article, I was going to use Gilmore' s1936 drawing of the reconstruction of *Apatosaurus louisae*, specifically the neck. But I decided that what would be better would be to articulated the individual cervical vertebrae, bone for bone. Even though sauropods are the longest, most difficult dinosaur to do research on, they have the more monographs done on them than any other group of dinosaurs. So it was easy to articulate *Apatosaurus lousiae*'s cervical vertebrae. When I started to draw the neck the vertebrae started straight out, but at the 8th and 7th vertebrae, the neck took a curve down, not up. The vertebrae has a natural curve down; i.e. the front of the vertebrae is actual lower then the back of the vertebrae, and gives the neck a natural curve downward (figure 4, 5).

I also drew *Diplodocus* and it did the same thin. If the neck was bent upward, it would either have to disarticulate at the "ball and socket" or it would have to have slid way past the postzygaphophysis (figure 6). I don't know how far the vertebrae could do this and is something I will look into, but when the neck is bent downward, it doesn't disarticulate. Stevens and Parish's research has also shown that the neck in *Apatosaurus* was more flexible than *Diplodocus*. *Apatosaurus* could turn its neck in a U shape, to look backwards and lift its head higher. *Diplodocus* couldn't' turn its head to look completely backwards, it could turn more of a "U", and less upward, but its neck was more flexible downward than upward. I would imagine that *Barosaurus* would be even less so. Why? It would have to do with how long each cervical vertebra is. *Apatosaurus* has the shortest, and *Barosaurus* had the longest. In *Brachiosaurus* the neck is shown at much lesser angle and upward. Quite unlike the 'S' shape Greg Paul gives them. This is how Sylvia Czerkas and Stephen Czerkas has *Brachiosaurus* in their Dinosaurs, a Global View. In Greg Paul's defense, here is his comment on the article that he posted on the Dinosaur Mailing List.

"These are some comments on the interesting article on using computers to simulate sauropods in the November Discover. It certainly is correct that diplodocid necks were held sub-horizontal, although it is not ever possible to tell exactly what the angle was because animals do not always habitually hold their necks in the neutral posture indicated by the cervical articulations. It is hardly likely that sauropods wandered about with their heads at ground level, they would tend to be surprised a lot. Nor were diplodocids regularly feeding on ground cover despite their low neck carriage. This is only thing we know because Fiorillo had SEM scan their teeth, and they show very little of the grit wear that occurs when animals eat dirty ground plants. In fact, the teeth show less wear than those of camarasaurs, which were feeding somewhat lower because they were shorter. *Camarasaurus* could barley reach the ground because they were short necks were so sharply flexed upwards at the neck base, and they could probably rear up as well. That diplodocids kept their teeth so clean suggests that they were feeding even higher on the well washed leaves of very tall plants by rearing. I have compared the strength of the dorsal vertebrae and find those of sauropods to be much larger and stronger than those giant mammals of the same mass. Therefore, the former were overbuilt for being just quadrupedal, and they should have been able to suspend their bodies from their hips more easily than the latter"

"The one thing I must disagree with is the sub-horizontal neck of *Brachiosaurus*. The problem is that only the lower halves of the vertebra at the neck-trunk juncture are known, the arches and the all important dorsal articulations are missing. So I do not see how it was possible to restore the posture of the neck base. I would not be at all surprised that a vertical swan neck is not possible, but the brachiosaurs were so high shouldered, and had true withers like giraffes, suggests their necks could be held at a steep angle. Likewise, brachiosaurus should have been able to reach ground level in order to drink, albeit just barely."

1

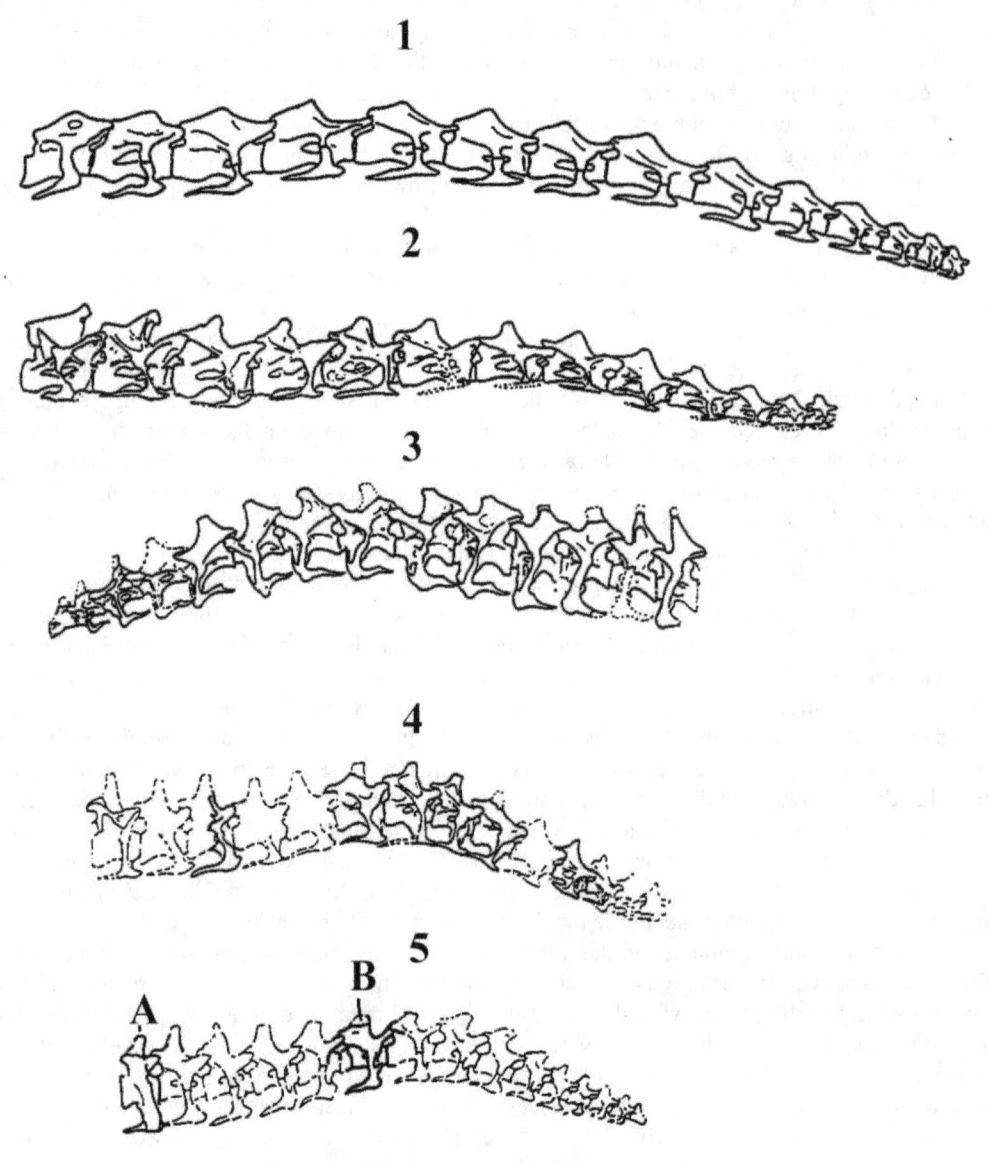

2

3

4

5

Figure 3), showing the different neck lengths of Diplodocids with the last dorsal 'cup' the same size. 1) *Barosaurus*; 2) *Diplodocus*; 3) *Apatosaurus lousiae;* 4) *Apatosaurus excelsus*; 5) *Apatosaurus ajax* (the last dorsal is of *Apatosaurus laticollis* (A) and dorsal 10 is *Apatosaurus immanis* (B), all part of the holotype of *Apatosaurus ajax.*

"Now, what I would like to see are computer simulations of the daily heat-storage/nocturnal cooling cycle of sauropods…"

Greg Paul is right in that the teeth should be taken into account, but there is much speculation on what the food was for each sauropod. Diplodocids are thought to be low browser, but some, thus a lower neck is required. But Greg Paul and others believe they were a high browser. Whether it was a 'ground 'feeder or a low tree feeder as opposed to high browser standing on its hind legs is debatable. I don't like a tripodal standing sauropod feeding. Recent articles have questioned whether or not diplodocids and a few other sauropods, could actually walk bipedally for awhile, but that is a discussion for another time. He is also right that no complete brachiosaurid neck is known, so how the neck was held at the last cervical and first dorsal is questionable. The neck of *Brachiosaurus* in my opinion didn't have that 'S' vertical neck, but it would appear to have been held, at the most, at the same angle, or just lower than that of a giraffe. One of the things that artists get wrong about brachiosaurids, and *Brachiosaurus* in particular, is that it is actually a slender animal. The neck and legs are slender; not robust as many artists make them.

Stevens and Parrish's research showed that the neck has a gentler slope in Camarsaurids but have not addressed euhelopids and *Omeiosaurus* yet. They have a "taller" neck, mainly due to dorsals 1 and 2 having a much steeper upward angle.

Stevens and Parrish are not the first ones to state that sauropod neck were not very flexible and were held nearly straight. J. Martin gave a talk and wrote a paper given at the 1987 Mesozoic Terrestrial Ecosystems symposium on *Cetiosaurus* that was mounted in the museum. They physically moved each vertebrae and came up with the neck couldn't move much. Also, as said before, Sylvia and Steven Czerkas also came up with this. Ken Carpenter while putting up the *Diplodocus* at the Denver Museum of Natural History also physically moved each vertebrae to see just how far it could move.

A few comments on *Amargasaurus*. The neck of *Amargasaurus* isn't that much of a problem as some may think. I talked to Rodolfo Coria about the long cervical neural spines. He doesn't believe they had a 'sail' neck. The cervical neural spines are not parallel to each other. One side is either higher or lower than the other. Also, the first few spines face backwards, while the last ones are more vertical. This was probably true for the how vertebral column free standing spines, not a sail (figure 6).

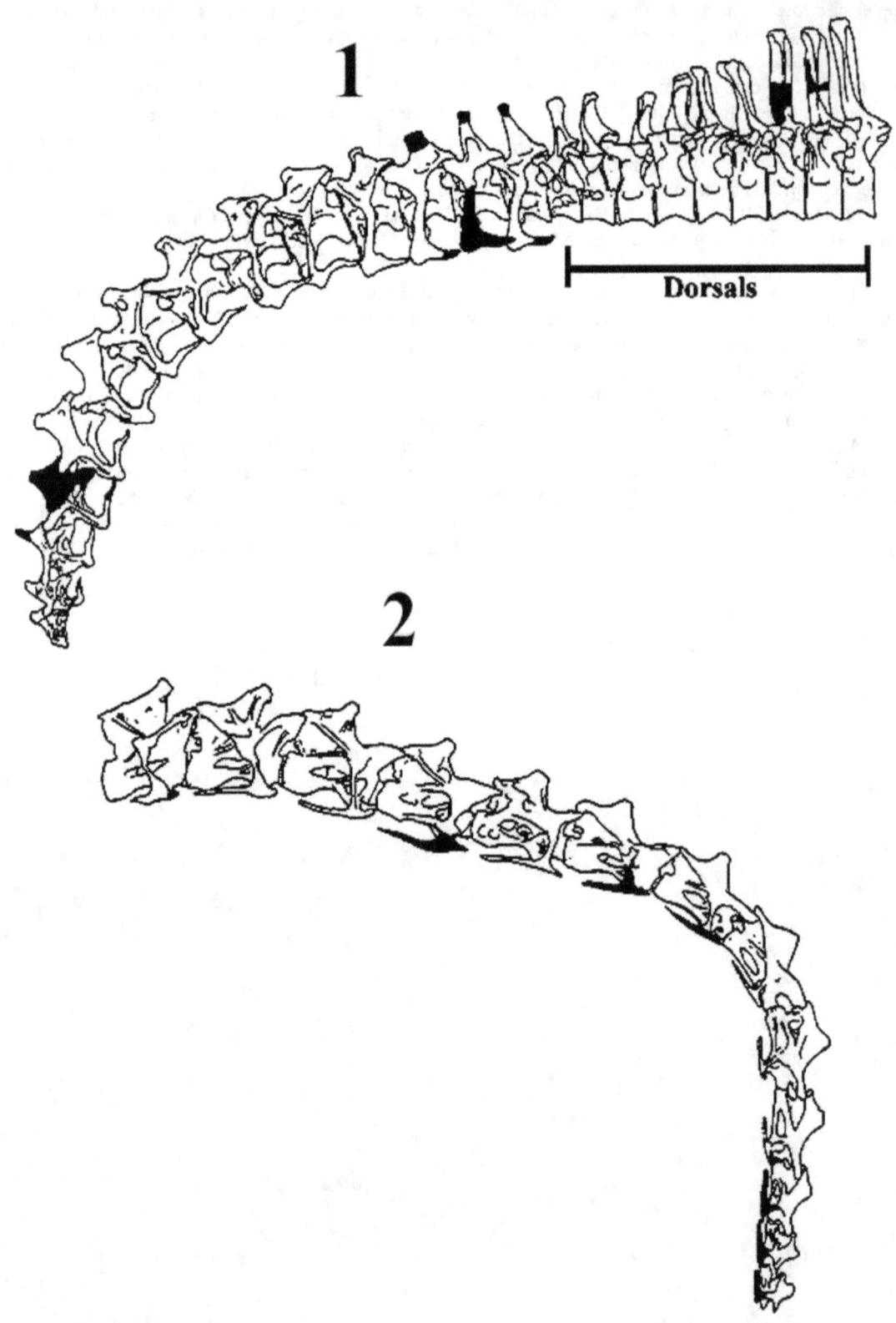

Figure 4), 'Natural' downward curve of the neck of 1) *Apatosaurus lousiae*, and 2) *Diplodocus carnegii*.

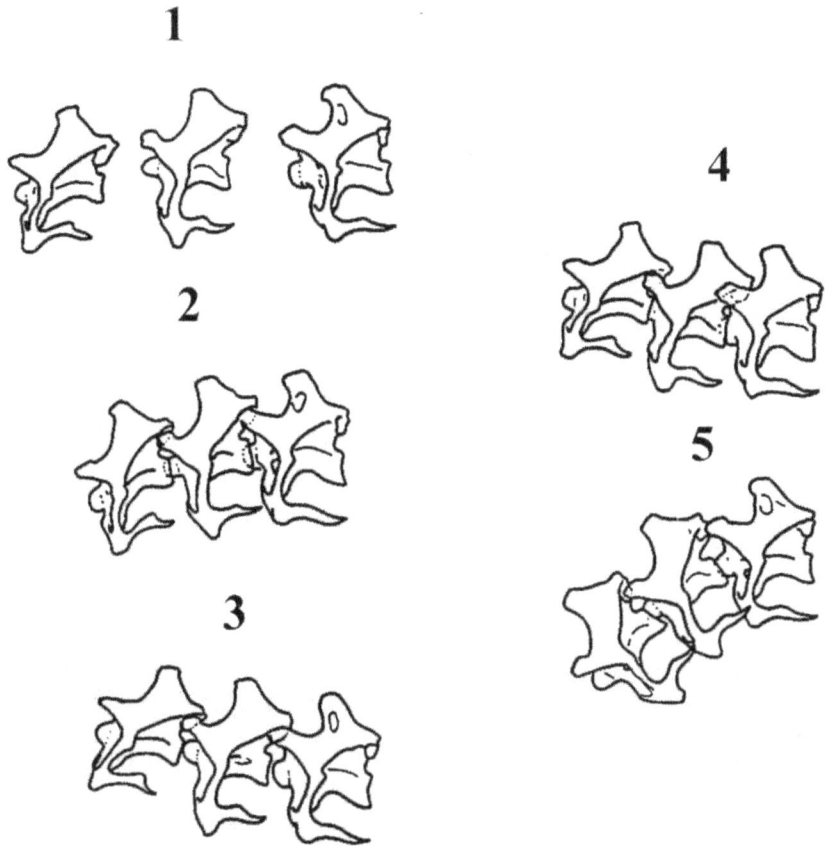

Figure 5), Dorsals 6-8 of *Apatosaurus lousiae* showing the natural downward curve of those vertebrae. Notice how the front end is lower than the back end. 1) All three vertebrae disarticulated; 2) All there in a level neck; 3) All three vertebrae tilting upward with out detaching the prezygopophysis with the postzygopophysis; 4) A downward tilt of the neck with nothing being disarticulated.

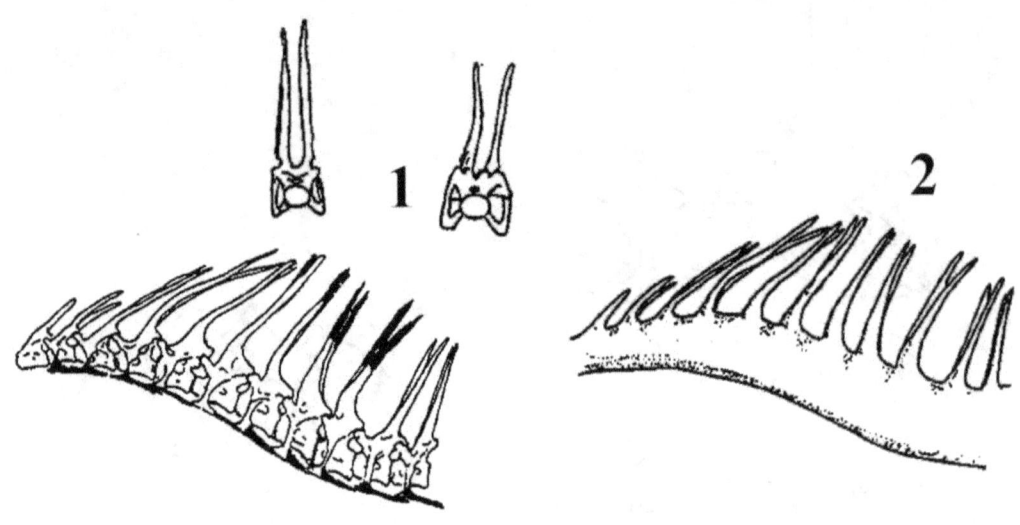

Figure 6), *Amargasaurus cazaui*, 1) with the articulated cervicals; and 2) fleshed-out neck.

Bibliography

Czerkas, S. J., and Czerkas, S. A., 1990, Dinosaurs a Global View: Dragons' World, 247pp.

Gilmore, C. W., 1936, Osteology of *Apatosaurus*, with special reference to specimens in the Carnegie Museum: Memories of the Carnegie Museum, v. 11, n. 4, p. 175-300.

Hatcher, J. B., 1904-1906, Osteology of *Haplocanthosaurus*: Memoires of the Carnegie Museum, v. 2, p. 1-75.

Janensch, W., 1950, Die Wirbelsaule von *Brachiosaurus brancai*: Palaeontographica Supplement n. 7, teil 3, lief. 2, p. 27-93.

Lull, R. S., 1919, The Sauropod Dinosaur *Barosaurus* MARSH: Memoirs of the Connecticut Academy of Arts and Sciences, v. 6, p. 5-42.

Martin, J., 1987, Mobility and feeding of *Cetiosaurus* (saurischia, sauropoda) - why the long neck?: In: Fourth Symposium on Mesozoic Terrestrial Ecosystems, short papers, edited by Currie, P. J., and Koster, E. H., p. 150-155.

Maxwell, W. D., Hallas, B. H., and Horner, J. R., 1997, Further neonate dinosaurian remains and dinosaurian eggshell from the Cloverly Formation of Montana: Journal of Vertebrate Paleontology, v. 17, supplement to n. 3, Abstracts of Papers, Fifty-seventh Annual Meeting Society of Vertebrate Paleontology, Field Museum, Chicago, Illinois, October 8-11, p. 63a.

Olshevsky, G., 1992, A Revision of the Parainfraclass Archosauria Cope, 1869, Excluding the Advanced Crocodylia: Mesozoic Meanderings n. 2, second printing, 268pp.

Osborn, H. F., and Mook, C. C., 1921, *Camarasaurus*, *Amphicoelias*, and other Sauropods of Cope: Memoris of the American Museum of Natural History, new series, v. 3, part 3, p. 249-387.

Ostrom, J. H., and McIntosh, J. S., 1966, Marsh's Dinosaur, the Collections from Como Bluff: Yale University Press, 388pp.

Salgado, L., and Bonaparte, J. F., 1991, Un nuevo Sauropodo Dicraeosauridae, *Amaragasaurus cazui* Gen. et sp. nov. De la Formacion La Amarga, Neocomiano de la Provincia del Neuquen, Argentina: Ameghiniana, v. 28, n. 3-4, p. 333-346.

Stevens, K. A., and Parrish, J. M., 1996, Articulating three-dimensional computer models of Sauropod cervical vertebrae: Journal of Vertebrate Paleontology, v. 16, supplement to n. 3, Abstracts of Papers, Fifty-sixth Annual Meeting, Society of Vertebrate Paleontology, American Museum of Natural History, New York, New York, October 16-19, p. 67A.

Wiman, C., 1929, Die Kreide-Dinosaurier aus Shantung: Palaeontologica Sincia, series C, v. 6, p. 1-67.

Zimmer, C., 1997, Dinosaurs in Motion: Discover, v. 18, n. 11, p. 96-104, 109.

PreHistoric TIMES

For the Dinosaur Collector and Enthusiast

April-May 1998   No. 28/29

80 Page Giant Double Issue

WILLIAM STOUT–PART TWO

U.S. $5.95 • Canada $6.95

0 74470 92325 1   04>

"Big Cats"
Artist
Mauricio Anton

Louis Marx
Dinosaur
Toys

Aurora
Prehistoric
Scences
Model Kits

Ford, T. L., 1998, How to Draw Dinosaurs. Ankylosaurs, the living tanks: Prehistoric Times, n. 28/29, p. 14-15.

Chapter 8

Ankylosaurs, the living tanks

Ankylosaurs have often been referred to as living tanks. Their large, oval bodies and their dermal scutes (armor) make them a most formidable animal which brings images of a slowly pondering animal, thrashing through meadows of newly evolved flowering plants.

I find these dinosaurs fascinating, but I won't be talking about their amour for this article. The armor varies greatly among the different families. Also there are new specimens that will be described that have bearings on the amour of ankylosaurs; some had very little armor.

Ankylosaurus is the largest of the ankylosaurs, but it is one of the least well known. The type skull consists of the top portion, a few teeth, cervical, dorsal and caudal vertebrae, fragmentary sacrum, right scapulocoracoid, ribs, dermal armor, and many other fragments. A referred specimen is more complete; it is from this specimen that the skull and more of the skeleton is known. It was first thought to be related to *Stegosaurus* and the skeleton was drawn like a *Stegosaurus*. It wasn't until years later, and new discoveries of new genera of ankylosaurs that the true body plan was discovered.

When I was at the Mesa Southwest Museum's Russian Dinosaur exhibit I noticed a triangular plate on the lower jaw of *Tarchia* and *Saichania*. I asked Jim Kirkland about this and said all ankylosaurid ankylosaurs had this. I was unaware of that and wish to pass this bit of information on to the readers (figure 1). *Ankylosaurus* lower jaw scute can't be seen very well, but it is there. Ankylosaurs also have a dermal scute over the eyelid. Walter Coombs described two *Euoplocephalus* skulls that had this scute (figure 2).

Ankylosaurs are often drawn with a short neck, yet they had a medium length neck. The body is very wide and short (in height), when looking from above it has an almost oval shape. The ilium is widest at the cranial (front) end and thinner at the caudal (rear) end. Looking at the body from the front, it looks like an overstuffed cow (figure 3). The body itself is longer than most illustrator's make it and I disagree with Ken Carpenter's interpretation that the body was short (figure 4). The legs were short, with *Talarurus* being the shortest. The hind legs had very massive muscles on the front of the leg due to the longer front portion of the ilium, and illustrators should take note of this (figure 5).

The front feet had 5 toes with hoof like claws on digits 1-3, while the hind foot had 3 hoofed clawed toes.

The tail dips downward from the ilium then levels out. The tail was most flexible at the base while from the middle to end of the tail was stiffer due to the transverseprocess (top portion of the vertebrae) on the caudals that connected one vertebra to the next, as well as being covered with tendons, which due to fossilization are called ossified tendons. This means from the mid portion to the club the tail didn't bend much (or possibly at all) and as stated before, it was the front of the tail that did all the moving. The tail acted more like a sledgehammer than a club.

The club itself varied form individual to individual and were not all the same size. There are several specimens of *Euoplocephalus*, some from about the same size individual, that have different size tail clubs. This is due to them either being a male or female or individual variation (figure 6).

Tony Thulborn had an interesting interpretation of the tail and that is it acted as a decoy, and the animal mimicked an ornithopod with the tail acting as the head and neck. This is not very well accepted theory.

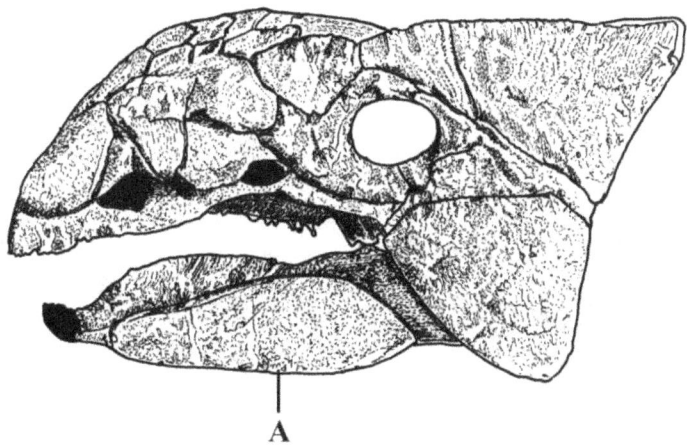

Figure 1), Skull of the referred specimen of *Ankylosaurus* with (A) showing the scute on the lower jaw (after Coombs, 1978). .

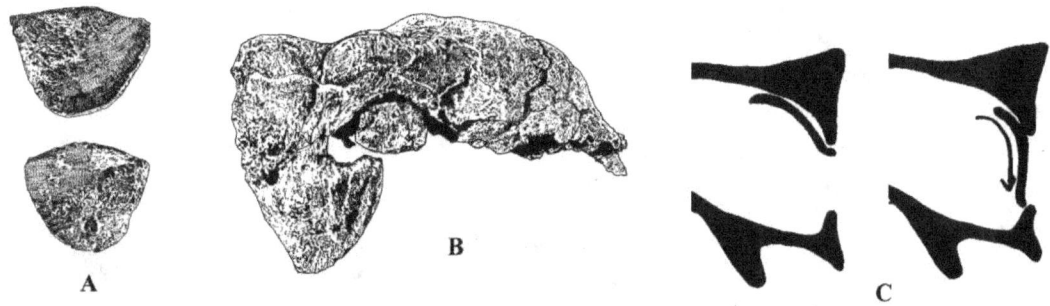

Figure 2), The boney eyelid of *Euoplocephalus tutus* (AMNH 5238), after Coombs, 1972); A) eyelid osteoderm; B) Side view of the fragmentary skull and C) Cross section showing how the eyelid closed (after Coombs, 1972).

Figure 3), Front view of the body of *Ankylosaurus*.

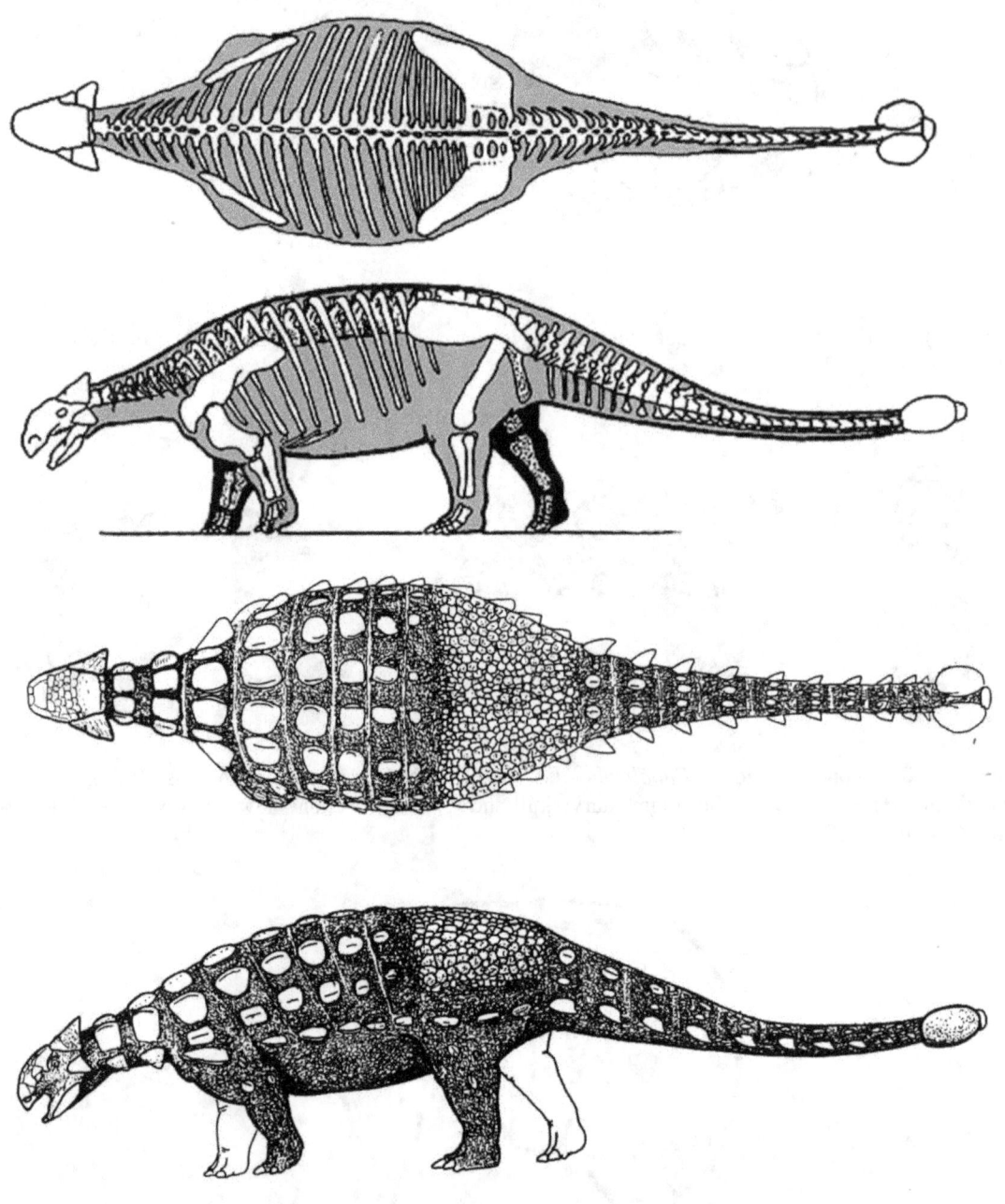

Figure 4), Modifed skeleton of *Ankylosaurus* from Coombs, (1979): **Editors note: I originally didn't put in the reconstructed animal, but I think its important enough to put it in now..**

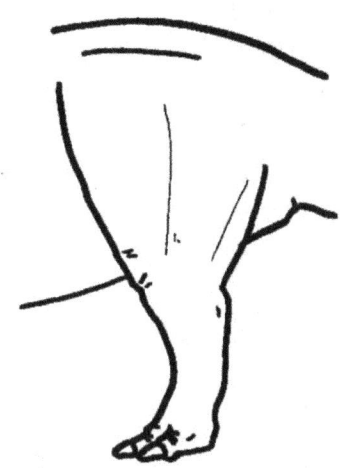

Figure 5), Side view of a hind leg of *Ankylosaurus* after Coombs, 1979.

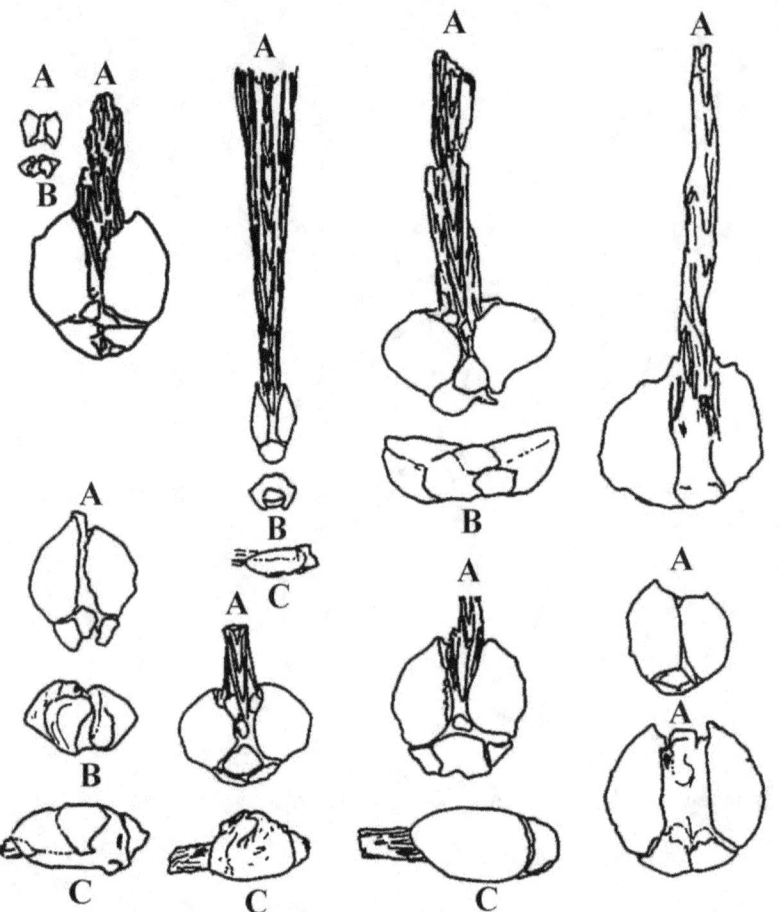

Figure 6), Several views of tail clubs of *Euoplocephalus* in A) top view, with some showing the ossified tendons; B) rear view, and C) side view (after Coombs, 1995).

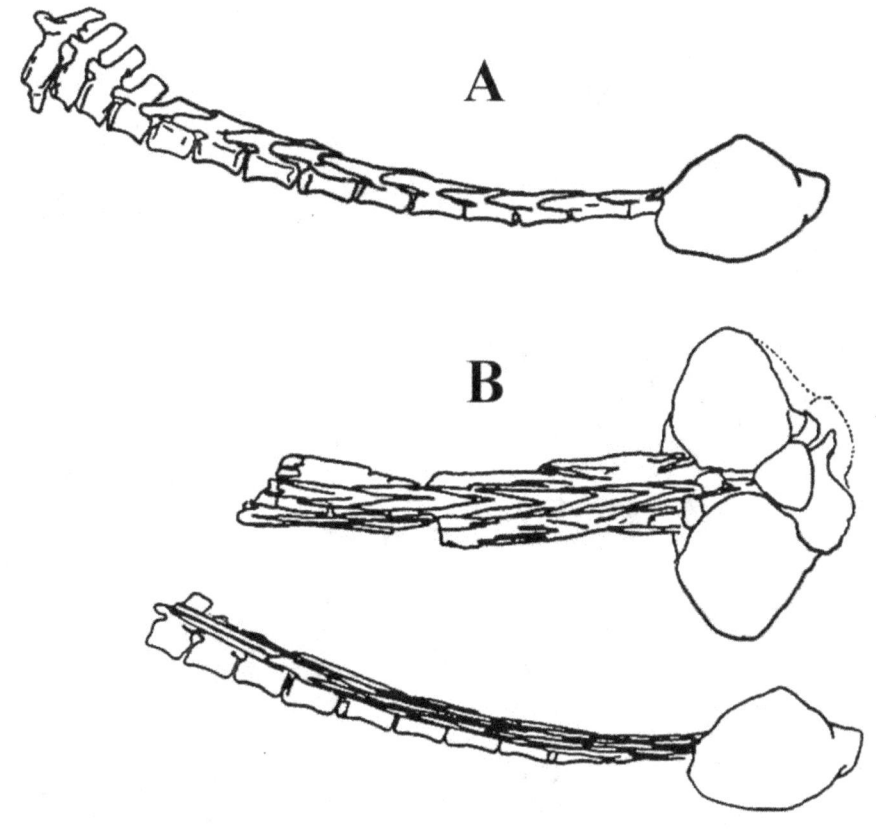

Figure 7), Side view of the tail of *Euoplocephalus* with (a) no ossified tendons, and B) with ossified tendons.

Bibliography

Carpenter, K., 1982, Skeletal and dermal armor reconstruction of *Euoplocephalus tutus* (Ornithischia: Ankylosauridae) from the Late Cretaceous Oldman Formation of Alberta: Canadian Journal of Earth Sciences, v. 19, p. 689-697.

Coombs, W. P. jr., 1972, The bony eyelid of *Euoplocephalus* (Reptilia, Ornithischia): Journal of Paleontology, v. 46, n. 5, p. 637-650.

Coombs, W. P. jr., 1978, The families of the ornithischian Dinosaur order Ankylosauria: Paleontology, v. 21, part 1, p. 143-170.

Coombs, W. P. jr., 1978, Forelimb muscles of the Ankylosauria (Reptilia, Ornithischia): Journal of Paleontology, v. 52, n. 3, p. 642-657.

Coombs, W. P. jr., 1979, Osteology and Myology of the hindlimb in the Ankylosauria (Reptilia, Ornithischia): Journal of Paleontology, v. 53, n. 3, p. 666-684.

Coombs, W. P. jr, 1995, Ankylosaurian tail clubs of middle Campanian to early Maastrichtian age from western North America, with description of a tiny club from Alberta and discussion of tail orientation and tail club function: Canadian Journal of Earth Sciences, v. 32, p. 902-912.

FOR THE DINOSAUR COLLECTOR AND ENTHUSIAST

PREHISTORIC TIMES

Dec. - Jan. 1997-98 • No. 27 $5.95

WILLIAM STOUT

U.S. $5.95 • Canada $6.95

10>

0 74470 92325 1

ALSO:
PALEONEWS '97
THE YEAR
IN REVIEW!

INTERVIEWS:
JOHN HORNER
ROBERT BAKKER
& MUCH MORE!

Ford, T. L., 1998, How to Draw Dinosaurs. The protoceratopians: Prehistoric Times, n. 30, p. 12-14.

Chapter 9

The Protoceratopians

Triceratops is the spotlight dinosaur this issue, but I'm not going to talk about those big rhino sized ceratopian's. instead I'll discuss their smaller cousins, the Protoceratopians. (You've no doubt noticed that I haven't put an 's' in Ceratop (s) ian. Why you may ask. Dinosaur expert, George Olshevsky told me that the proper way to say Ceratopsians is Ceratopian, which is why I'm using it that way).

The skeleton of *Protoceratops* is often mounted with low, bowed, and sprawling legs. This gives the animal a lower stance (figure 1A, B), and consequently they are drawing the same way. This persistence in drawing the animal in this stance is wrong. Protoceratopians had a normal high, ornithopodian upright stance. The hind legs are well under the body and the front leg slightly further out (figure 1C, D). V. S. Tereshchenko has written two articles on this, one in 1994, the other in 1996. His research showed that *Protoceratops* had a higher stance and was more active, agile animal.

Some protoceratopians are thought to have been totally bipedal, like *Microceratops* (figure 2) or partially bipedal like *Protoceratops* when running (according to Bakker, 1967, though this is not well accepted for *Protoceratops*). A new *Montanoceratops* skeleton is bipedally mounted in Jim Kirkland's Ceratopian exhibit. This specimen has short forelimbs and long hind limbs.

The front feet of protoceratopians have 5 short toes; with the first 3 having claws. The hind feet has long metatarsals, and 4 toes, with long claws. In *Protoceratops* these claws are blunt, while other protoceratopians have sharper claws (figure 3). The blunt claws in *Protoceratops* may have been used to dig burrows. Watabe et al., (1995, 1996, a joint paleontological team from Japan, America, and Mongolian Museums), has reported a nest of juvenile Protoceratopians that may have been in a burrow. The sedimentation supports their belief. Also, Jim Kirkland and Paul Johnson are currently working on a theory that the reason why some *Protoceratops* skeletons are found at an angle (and some sitting) is because they were buried in burrows much like today's groundhogs or prairie dogs.

Oddly the skull of *Protoceratops* is quite large. I've drawn *Leptoceratops* with a smaller skull than it had because it looked wrong, this in itself is wrong. I've redrawn it after looking at how Greg Paul drew it and realizes that he was right and I was wrong (figure 4). Adult male, female and juvenile *Protoceratops* are known. The males have a small bump on the nasals (this bump is bifurcated, with a small ridge along the middle. In the larger ceratopian p;ceratopians, this is a juvenile state) and large wide frill, while the female have no bump and a smaller frill, and juveniles have smaller frills (figure 5).

The tail of protoceratopians vary from genus to genus. In *Leptoceratops* the caudal neural spines are short. *Protoceratops* and *Montanoceratops* have high caudal neural spines. Some attribute this to a partially aquatic life style. The problem with this is that the high neural spines are in the middle of the tail and aquatic animals have tall caudals near the end of the tail, which would help in propulsion (figure 6).

The jaw muscles in ceratopains in general have been interrupted to connect to the top of the frill. The frill is not a strong solid, piece of bone. The bone is porous, with two large holes or fenestra and was not that strong, the animal would have broken its frill if the jaw muscles was attached to the frill. The proper position of the jaw muscles is low on the frill.

A few comments on the fighting *Protoceratops* and *Velociraptor*. The prevailing theory is that *Velociraptor* was attacking the nest of *Protoceratops*. Because the aggressive behavior attributed to theropods, it is only natural to assume that *Velociraptor* was attacking *Protoceratops*. But it is not unreasonable that *Protoceratops* was attacking the *Velociraptor's* nest, making *Protoceratops* the aggressor. Nor is it totally impossible that *Protoceratops* was an egg-eater. I can see it picking up an egg in its mouth and puncturing it with the two sets of two premaxillary teeth. That is not to say that it just ate eggs, but it could do so when the occasion arrive.

I will make one comment on *Triceratops*. This has to do more with reconstructing and drawing the skull of the subject animal correctly. When I was doing the illustrations for George Olshevsky's article on ceratopains for Gakken Mook's Dinosaur Frontline magazine I needed to draw *Triceratops flabellatus*. I was going to draw *Triceratops flabellatus* using Marsh's drawing of the skull. I came across a short article by Lull that was published as a supplement to his revision monograph on ceratopains in 1934. He had a photo of the skull of *Triceratops flabellatus* and it looked nothing like the drawing by Marsh. The skull is longer for one thing and the frill is different. It is very important to draw the bones correctly and I can not stress this enough. When I find out that I've draw a bone or skeleton wrong I go back and correct it. From now on when you see the drawing by Marsh or a redrawing of *Triceratops flabellatus* using Marsh's drawing you'll know that the author didn't research that drawing and that it is wrong (figure 7).

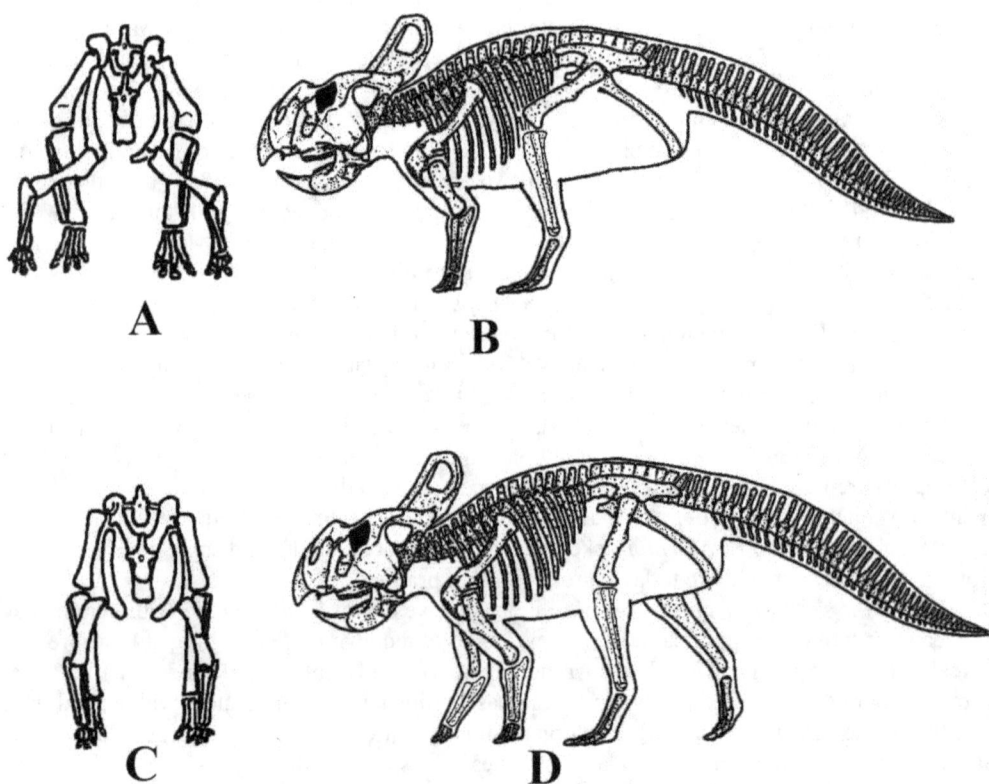

Figure 1), *Protoceratops* skeleton; A) Front view of the skeleton showing a sprawling stance; B) Side view of the skeleton; C) Front view of the upright stance; D) side view of the skeleton.

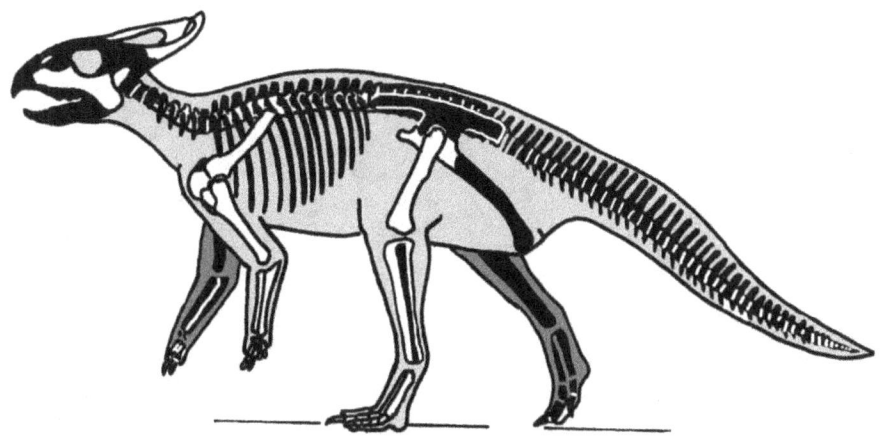

Figure 2), *Microceratops* skeleton.

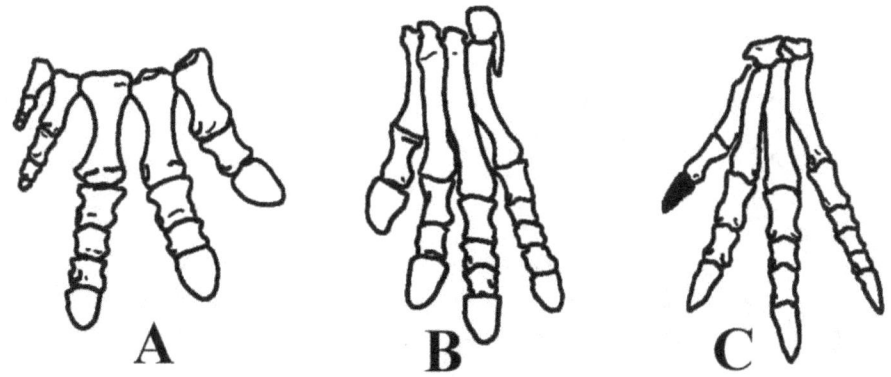

Figure 3), Front foot of A) *Protoceratops*; B) Hind foot of *Protoceratops*; and C) Hind foot of an unnamed, fragmentary Protoceratopian specimen from the Two Medicine Formation (USNM 13863) (**Editors note: USNM 13863 has since been referred to *Cerasinops hodgskissi* by Chinnery & Horner (2007)**).

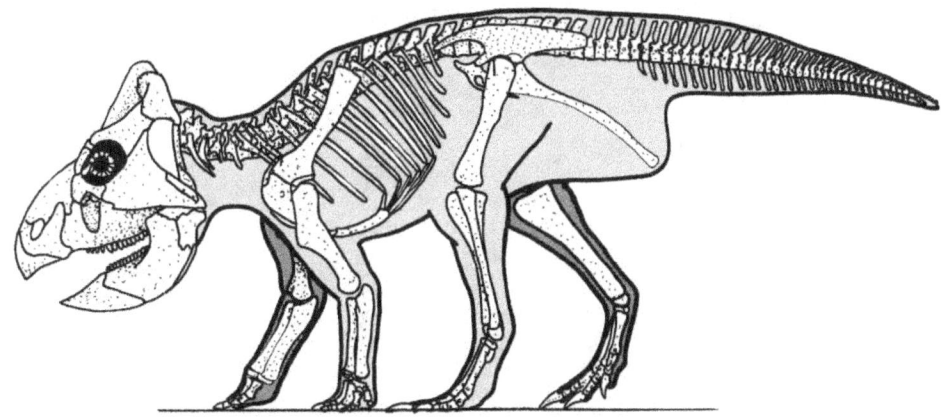

Figure 4), *Leptoceratops* skeleton.

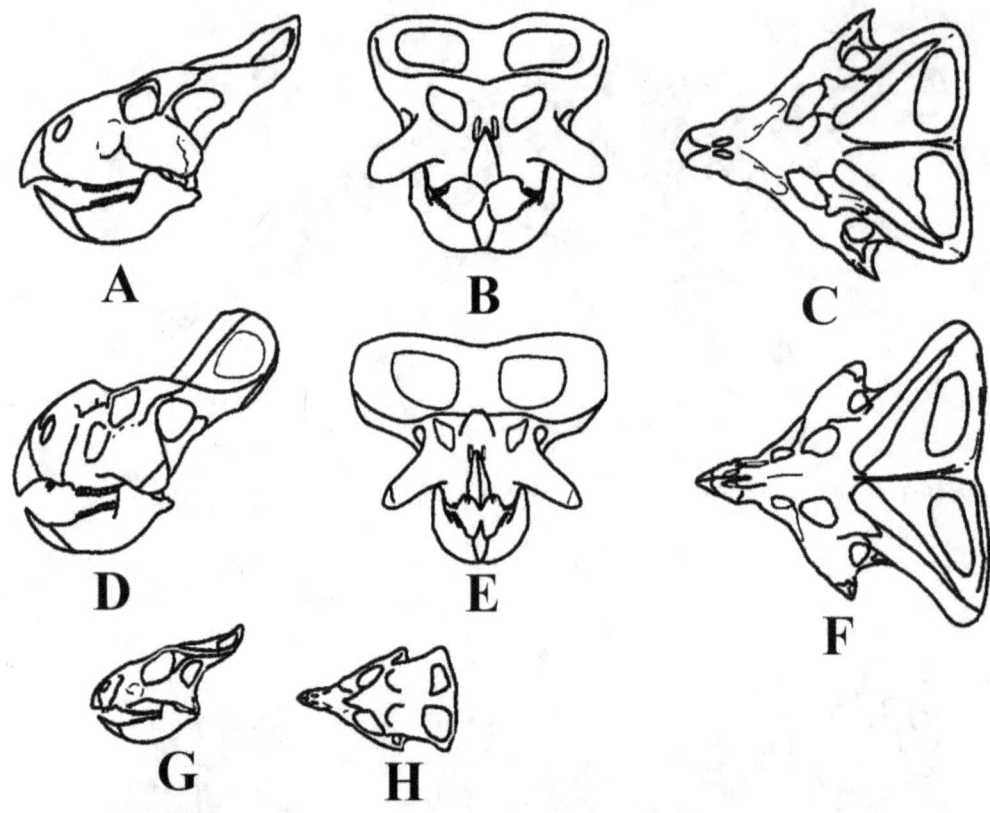

Figure 5), *Protoceratops* skulls; A) Side view of a female; B) Front view of a female, C) Dorsal view of a female; D) Side view of a male; E) Front view of a male, F) Dorsal view of a male; G) Side view of a juvenile; and H) dorsal view of a juvenile.

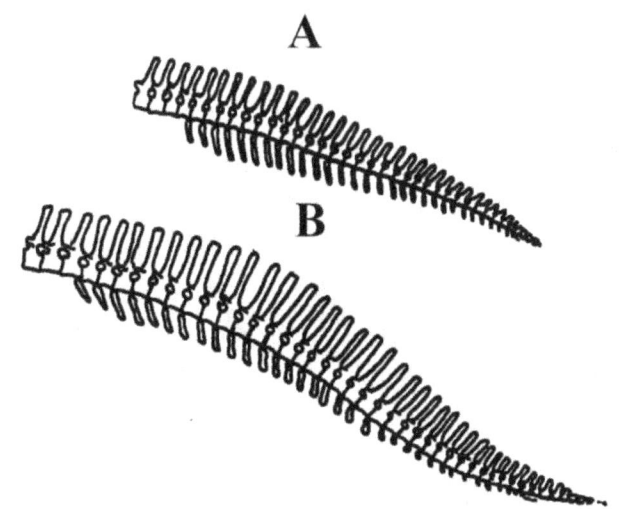

Figure 6), A) *Leptoceratops* tail; B) *Protoceratops* tail.

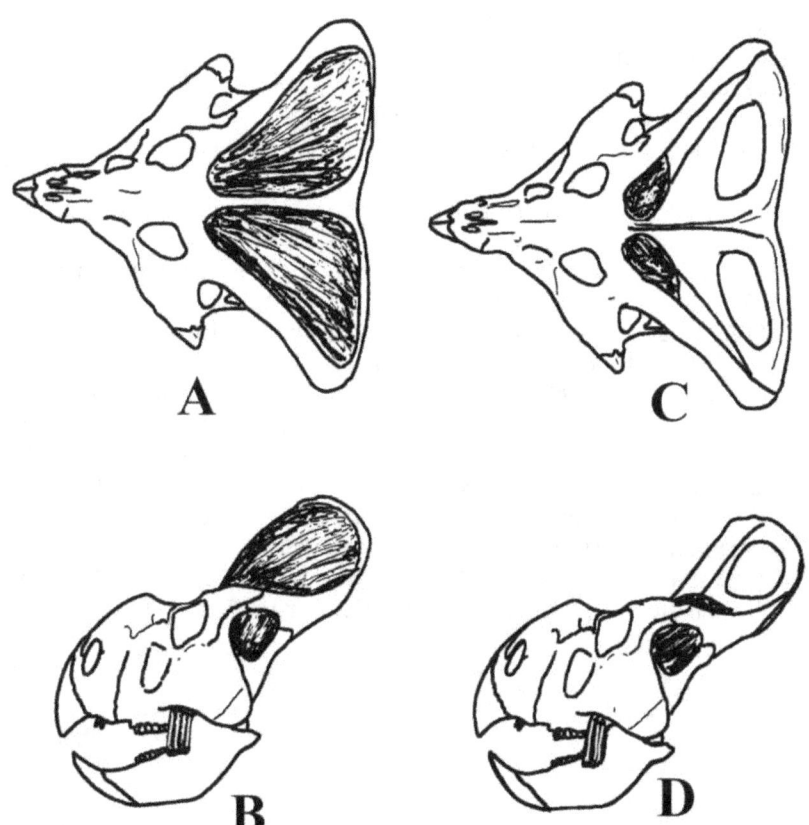

Figure 7), A) Top view of the skull of *Protoceratops* showing the improper jaw muscles; B) Side view of the skull; C) Top view of the skull of *Protoceratops*, showing proper position of the jaw muscles; D) Side view of the skull.

Figure 7), Side view of *Triceratops flabellatus*; A) according to Marsh, 1889; and B) Side view according to Lull, 1934.

Bibliography:

Brown, B. B., and Schlaikjer E. M., 1940, The structure and relationships of *Protoceratops*: Annual of the New York Academy of Science, v. 11, article 3, p. 133-266.

Lull, R. S., 1934, Skull of *Triceratops flabellatus* recently mounted at Yale: The American Journal of Science, 5th series, v. 28, p. 439-442.

Marsh, O. C., 1889, Notice of Giant horned Dinosauria from the Cretaceous: American Journal of Science, 3rd series, v. 38, p. 173-175.

Tereshchenko, V. S., 1994, A reconstruction of the erect posture of *Protoceratops*: Paleontological Journal, v. 28, n. 1, p. 104-119.

Tereshchenko, V. S., 1996, A Reconstruction of the Locomotion of *Protoceratops*: Paleontological Journal, v. 30, n. 2, p. 232-245.

Watabe, M., Suzuki, S., Tsogtbaatar, K., and Barsbold, R., 1995, Results of the Hayashibara Museum of Natural Sciences and Geological Institute, Academy of Science of Mongolia, Joint Paleontological Expedition in the Gobi Desert: 1994 and 1995: Journal of Vertebrate Paleontology, v. 15, supplement to n. 3, Abstracts of Papers, Fifty-Fifth Annual Meeting Society of Vertebrate Paleontology, Carnegie Museum of Natural History, Pittsburgh, Pennsylvania, November 1-4, p. 57a.

Watabe, M., Suzuki, S., Tsogtbaatar, K., Barsbold, R., and Weishampel, D. B., 1996, Hatchlings of the dinosaur *Protoceratops* (Ornithischia, Ceratopsia) from the Upper Cretaceous Locality, Tugrikin Shire, Gobi Desert, Mongolia: Journal of Vertebrate Paleontology, v. 16, supplement to n. 3, Abstracts of Papers, Fifty-sixth Annual Meeting, Society of Vertebrate Paleontology, American Museum of Natural History, New York, New York, October 16-19, p. 71A.

DINOSAUR WORLD

Vol. 1 no. 3   *October 1997* $ 7.50

CERATOPSIAN SPECIAL ISSUE!

Ford, T. L., 1997, Ceratopsian stance. Dinosaur World, v. 1, n. 3, p. 12-17.

Chapter 10

CERATOPIAN STANCE

How ceratopians stood has been hotly debated for decades. Was the "Russell Sprawl" (coined after Lois Russell who first wrote about the sprawling stance) which is supported by Johnson and Ostrom, or a "Rhino dance" that Bob Bakker and Greg Paul subscribe to, evident in ceratopians? Which view is correct? Neither, both or something in-between? Opinions concerning the stance of ceratopians have been based mainly on the pectoral girdle, and skull remains. But the entire skeleton should also be considered. In attempting to resolve this longstanding controversy, let's examine all perspectives.

For their interpretation, Johnson and Ostrom relied on a model of the pectoral girdle of a recently found *Torosaurus latus* skeleton that was found in the Hell Creek Badlands of northeastern Montana. The fragmentary skull and partial skeleton reside in the Milwaukee Public Museum MPM VP6841. In their 1995 paper on the stance of *Torosaurus*, the authors stated that the skull of *Torosaurus* was 1/3 the weight of the whole animal.

But is this true? (figure 1) The answer significantly impacts our perception of ceratopsian stance.

CERATOPSIAN HEADS

The skull in dinosaurs can be a very misleading. Skulls sometimes appear to be very tall and thick, but they are actually quite 'hollow'. It is easy to see how hollow a theropod skull is with the fenestrae visible, but it isn't so obvious in the case of a ceratopian. The skulls of ceratopians are also very hollow (even neglecting the fenestrae). For example, examine my "cut-away" view down the mid-line of a *Torosaurus* skull. (Authors note: Even though no *Torosaurus* skull has been sectioned, for illustrative purposes, I've modified *Triceratops flabellatus*, YPM 182, after Hatcher, Marsh & Lull, 1907 and Lull, 1933.) You can see just how hollow the skull was. The rostrum, and premaxilla area is the solid bone at the cranial end of the skull that connects the left and right side. The nasal, frontal, and prefrontal are what the top of the skull attaches to on the left and right sides, with the braincase and occipital being the only "solid" portions in the center of the skull.

An interesting point is that the brain of *Triceratops* and *Torosaurus* (?) and possibly all ceratopians actually was located beneath the eyes. A possible reason for this is postulated below.

The area around the premaxilla and braincase is hollowed out, in life possibly filled with muscles, sinuses or other soft tissue. There is no skeletal palate to speak of, but there had to have been a palate of some kind made up of cartilage to prevent the food from going into the naries area, along with a soft tissue element, perhaps a valve, to close off the air passage of the naries to the mouth while eating. In dinosaurs the naris goes into the top of the mouth, entering from either the middle or back of the mouth. (I am unaware if any may have had a second palate).

The maxilla and dentary are actually very narrow and their width has often been exaggerated in others' restorations. The lower jaw fits inside the upper jaw with the upper and lower teeth curving inward to each other, making a long 'shearing edge'. Ceratopians did not masticate like mammals, but 'Gansu' knifed their food with the jaws, cutting the plant material ever so finer.

The frill and horn cores are fairly porous. In some specimens the outside of the horn core has 'grooves', which are believed to represent channels for blood vessels which nourished the horn core sheath. The outside of the frill, in some specimens, also has grooves. Does this mean the frill was also covered with a horny covering? That may seem ludicrous, but it's worthy of further consideration. Keith Rigby Jr. believes that the porosity of the frill and horns was a thermoregulatory consequence.

There is a hollow space between the top of the brain case and the top of the skull. (figure 2). This is considered to be a sinus area, possibly filled with either 'air' or some sort of fluid. I've wondered about this area and what its true function was. Normally the 'frontal' (a set of bones normally at the very top of the skull) bones have fenestrae or holes at the back edge for the jaw muscles to attach to. The frontal is closed in *Torosaurus* and *Triceratops*. In ancestral ceratopians this area is not closed over.

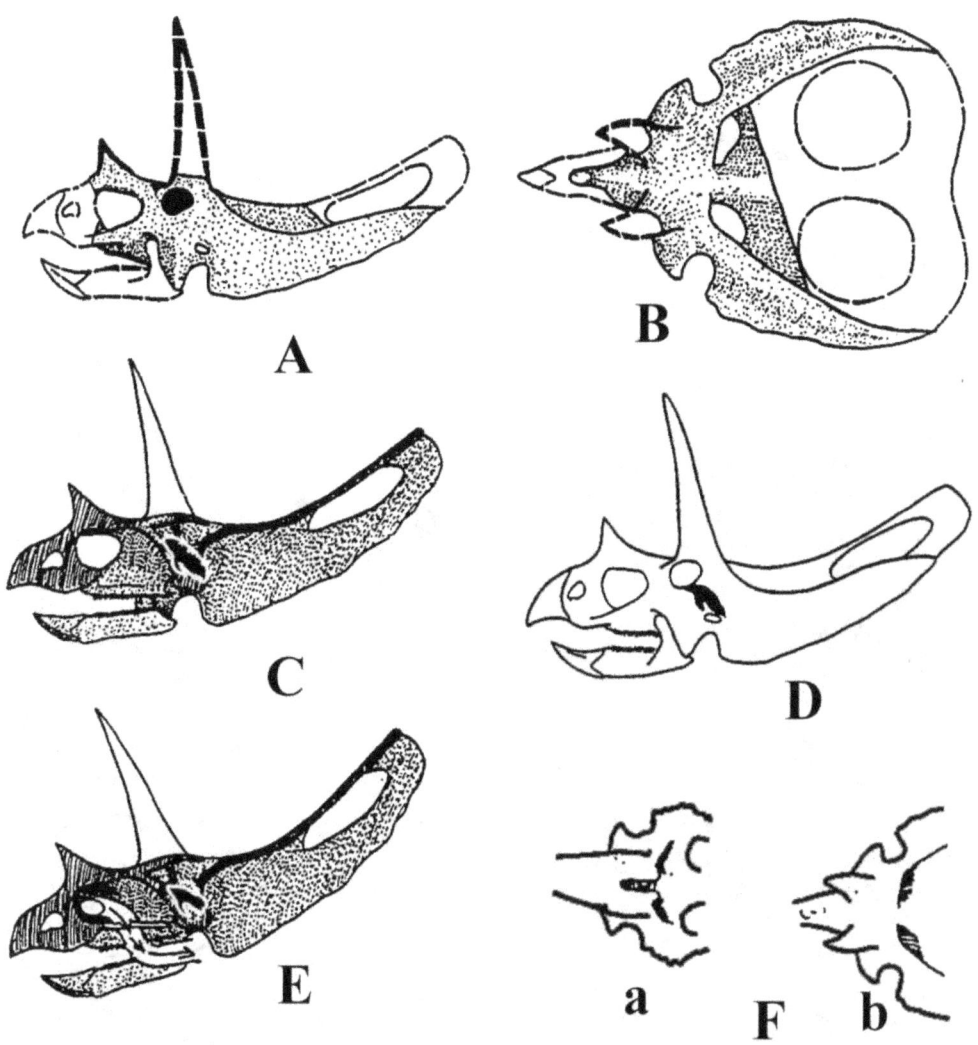

Figure 1). Skull of *Torosaurus latus* (modified from Marsh, 1896) in A). Side view, B). Top view, C). Cut away view showing the 'sinus' (a) and brain (darken in). D). Showing where the brain lies. E). Cut away view showing possible air path from the naries through the (a) soft palate. F). Top view of (a) *Styracosaurus parksi* showing the 'hole' in the top of the skull and the supertemporal fenestra, and (b) top view of the skull of *Torosaurus* showing the closed over 'hole' and the supertemporal fenestra.

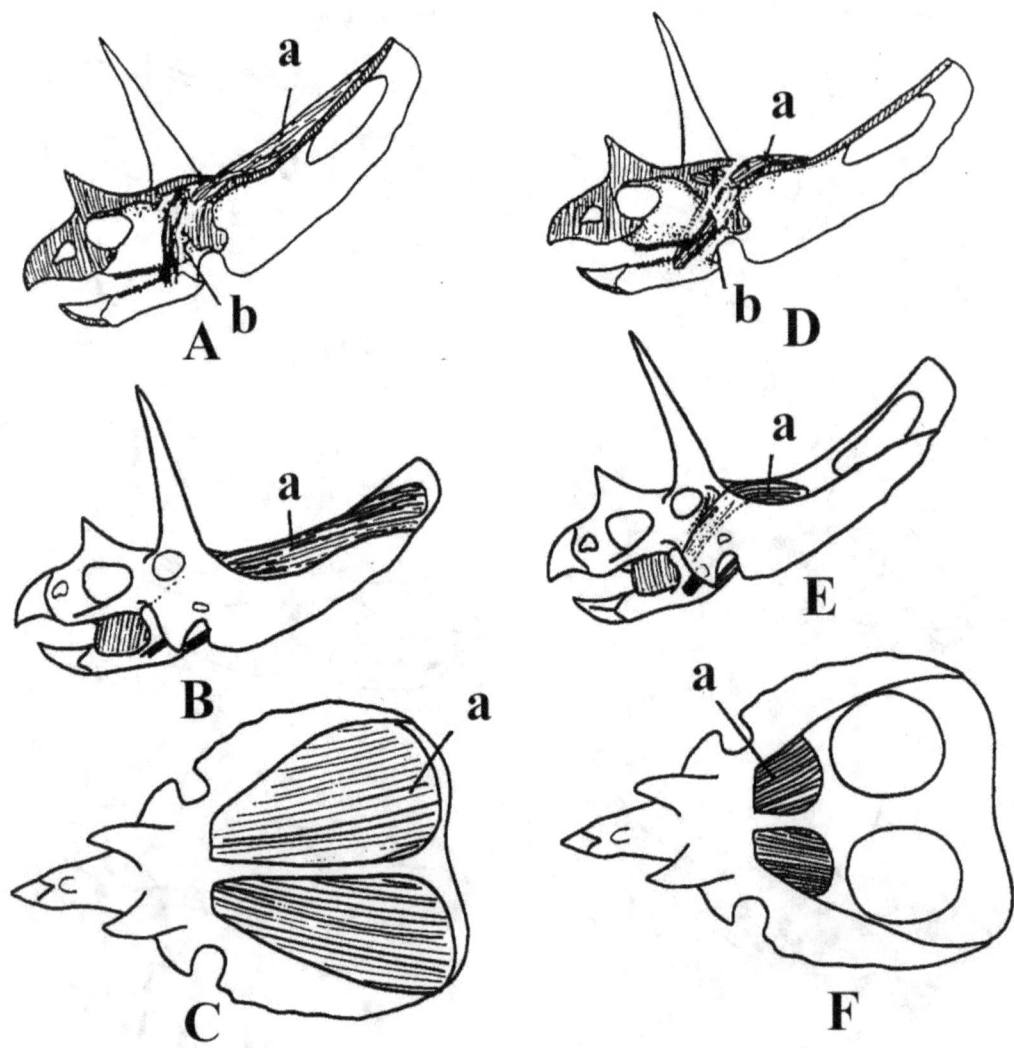

Figure 2). Skull of *Torosaurus latus* showing (a) pseudotemporalis muscle stretching to the back of the fenestra (you'll note how much this muscle stands above the frill, which would have made it vulnerable to attract) and (b) mandibulae externus superficalis atthces muscle attaching to the inside of the orbit (after Russell, 1935 and Ostrom 1964) in side view A, B, and top view C; Mandibulae externus superficalis atthces modified after M olnar 1973 showing the true (?) position of the pseudotemporalis muscle attaching to just beyond the supratemporal fenestra in side view D, E, and dorsal view F). (Note: I goofed on the marking the drawing, the left side is A, B, C the Right side is D, E, F).

While researching an article about theropod lips for the Mesa Southwest Museum, I encountered Ralph Molnar's (1973) thesis on the cranial morphology of *Tyrannosaurus*. He explicitly addressed all the muscles of the skull of *Tyrannosaurus rex*. In life, the pseudotemporalis muscle attached to the inside of the surangular, passed through the skull, and attached to the back side of the frontal through a fenestra. I imagine that the muscles of the jaws of ceratopians wouldn't functionally differ. How could the pseudotemporalis attach to the frontals if they were completely closed?

I imagine, (yet need to look at a three dimensional hollowed out skull - MRI image - eventually to check this), is that the top of the braincase is actually the frontal area, and the pseudotemporalis fits, inside the 'sinus'. Ostrom (1964) shows the pseudotemporalis fitting differently than in Molnar's thesis. He shows the pseudotemporalis attaching to the backside of the orbit. I believe that there was tendency in derived species for the top of the skull to become closed over the fenestra and over the pseudotemporalis. Perhaps an underlying reason for this is that the pseudotemporalis was so large that in its connection to the rear of the frontal bone would not interfere with the eyes.

Another muscle to consider is the adductor mandibular externus superficalis atthces (Temporalis of Russell, 1935). This muscle attached to the outside of the surangular then passes on the inside the skull, and out the supratemporal fenestra and attached to the backside of the skull, or on the parietal. In a 1935 paper Loris Russell shows several ceratopian taxa with this muscle attaching to the back of the frill, covering the large fenestra in the middle of the frill. This would not only add weight to the skull via soft tissue, but the 'force' exerted in contracting the muscle would, in my opinion, have damaged the frill. The supratemporal fenestra is shallow, allowing only the 'end' of the muscle to fit through, and this muscle attached to a point just beyond the fenestra and near the beginning of the parietal.

There is also a ridge of bone on the parietal which bisects the rear of the skull. Peter Dodson also believes that the jaw muscles attached to the parietal. In the frill of *Torosaurus*, there are two large fenestrae (as in nearly all the ceratopians except *Avaceratops*, *Diceratops* [now *Nedoceratops*] and *Triceratops*). What purpose did the fenestrae fill has been debated and is still in need of resolution. The muscles WOULD have added weight to the skull as 'wet tissue', but with the adjusted contribution of muscle added to the skull mass, it would not have exceeded 1/3 the total body weight. Also the horn cores would have added "hard tissue' weight. However, there is insufficient additional weight as to suggest a sprawling stance.

POST CRANIAL REMAINS

The first 2 to 3 cervical vertebrae are fused together forming a solid base for muscle attachment (figure 3). (This arrangement varies within the genera of ceratopians and incorporates the atlas and axis.) The neck muscles were attached to the back of the frill and the cervical vertebrae. They were not attached to the tip of the underside of the frill, but just behind the frill. The vertebrae itself is "amphiplatyan", meaning both sides of the vertebrae are 'flat'. There was cartilage between the vertebrae. Amphiplatyan vertebrae would have increased the mobility of vertebrae. However, several ceratopian skeletons have been found with ossified tendons running along the mid to caudal dorsal vertebrae, over the pelvic girdle and crainally to some caudal vertebrae. The tendons would be able to move slightly, but the main effect would be to stiffen/strengthen the back. The transverse processes (those two pieces of bone that stick out like a cross on the vertebrae) curve upward (in a plane). The vertebral column itself has a slight bend upward from the cervicals to the mid dorsals. The cranial dorsal ribs were rather straight, with subsequent ribs expanding outward. The front of the pectoral is narrow, and rearward is barrel-shaped. The cervical end of the rib cage was narrow, while the caudal end had a more rounded shape.

LIMB & PECTORAL ELEMENTS

The pectoral girdle consists of the scapulae, coracoids, and sternum. (figure 4). The glenoid is where the scapula and coracoid meet, and where the humerus attaches. The glenoid faced down and outward with no part of the glenoid on the inner surface. The scapula-coracoid contact lies below the vertebral column and about 1/3 third the length of the pectoral girdle in front of the first dorsal ribs. From the front of the pectoral girdle, the scapula looks like a thin straight line, with the coracoids curving inward meeting the sternum at the lower edge of the coracoid, giving the pectoral girdle a 'U' shape. The sternums (both left and right) were a rectangular bone that attached to each other along the mid-line.

The humerus of ceratopians (Fig.4) is a thick heavy bone. The humeral head is on the top upper side of the humerus. Also the 'head' is offset with some of the humerus on the inside. The ulna condyle lies in a straight line with the humeral head. The humerus is strongly curved between the posterior and anterior end, and slightly twists outward with the ulnae and radial condyles angled away from the center. The radius condyle lies on the outside of the humerus, and the ulna condyle on the inside. The radius and ulna only slightly cross each other.

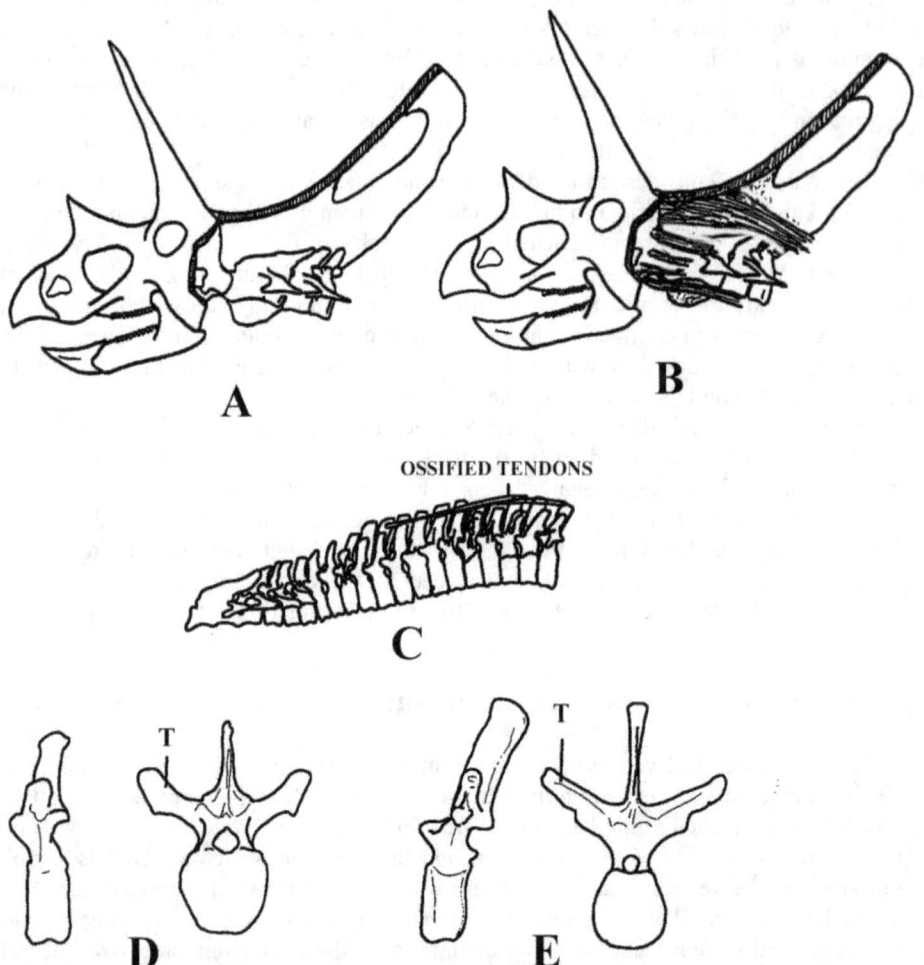

Figure 3). A). Skull of *Torosaurus latus* showing position of the fused cervical vertebrae. B). Showing the neck muscles of *Torosaurus latus*. C). Fused cervical and cervical and dorsal vertebrae of *Triceratops brevistrostris* (BSP 1964 I 458) showing the natural curve of the back and also the ossified tendons; D) Anterior dorsal vertebrae of *Triceratops prorsus* USNM 4842 (after Marsh) in (a) laterial and (b) anterior view; E), Posterior dorsal vertebrae of *Triceratops prorsus* USNM 4842 (after Marsh) in (a) lateral and (b) anterior view, showing the position of the transverse processes.

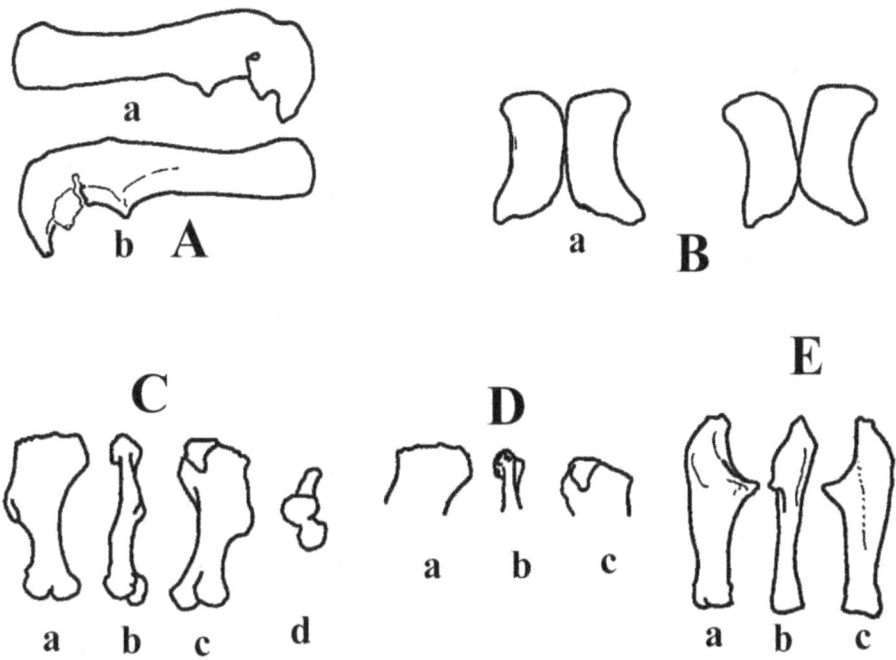

Figure 4). A). Scapula and coracoid of MPM VP6841 in a, laterial and b, medial view showing position of the glenoid (after Johnson and Ostrom, 1995). B). Sternum after Brown 1906 of Triceratops sp (AMNH 971), (a) modified after Johnson and Ostrom, 1995 (b). C). Humerus of MPM VP6841, in (a), ventral, (b), lateral, (c), dorsal and (d), distal views (after Johnson and Ostrom, 1995). D) Top half of humerus showing the position of the humeral 'head' (a). E). Ulna of MPM VP6841 showing the olecranon, in (a), anterior, (b), medial, and (c), posterior position (a'). Note: in (a), ventral, (b), lateral, (c), dorsal and (d), distal views Goes straight across, as does the ulna. Forgot to mark that.

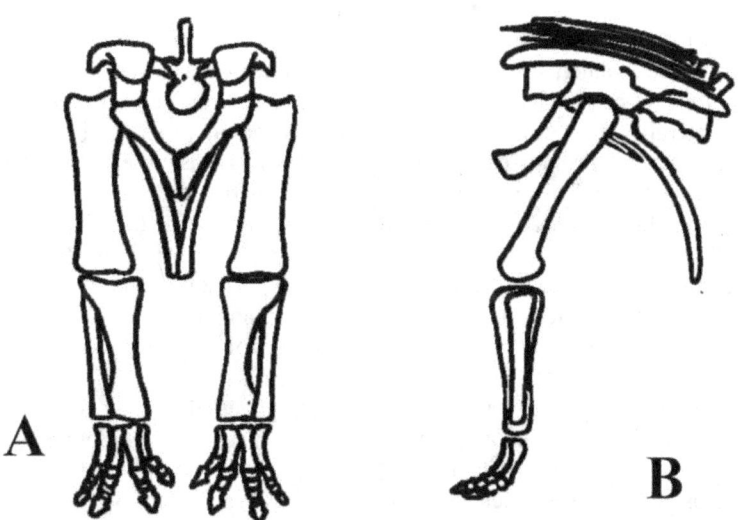

Figure 5). Pelvic girdle of *Torosaurus* in front (a) and lateral (b) views showing the erect position of the hind legs. Note no pelvic girdle of *Torosaurus* is known so I've modified the pelvic and hind legs from *Triceratops*.

The ulna has a large crest ('funny bone' or the olecranon). Sprawling animals have short olecranon. Why? A large olecranon would have hindered its sprawling or horizontal movement of the humerus, while mammals, with an erect gait, have a large olecranon. Fully quadrupedal dinosaurs, stegosaurs, ankylosaurs and ceratopians all have a large olecranon.

Were stegosaurs and ankylosaurs sprawling animals? No. If the ulna bears a large olecranon then it has (or had) an erect gait. The ulna and radius would fit on the 'ball' of the condyles of the humerus and not in between. The radius (not shown) is a straight bone.

Hunt, et al. (1997) has recently written an abstract about pelycosaur prints from New Mexico. They noted that the gauge of the prints was narrow, and there was no tail drag. The conclusion is that pelycosaurs had a narrow erect gait, and erect stance. Pelycosaurs do HAVE a large olecranon, which fortifies the anatomical "rule".

Because the ulna slightly twists outward distally, the feet have a slight twist outward as can be seen in the trackways of *Tetrapodosaurus*.

There is no controversy about the position of the hind leg. It was erect, which is typical of all ornithopods. (figure 5).

STANCE

What does all of this have to do with the stance? Consider the sternum. There are both a right and left sternum. They were connected to each other by ligaments and muscles along with muscles from the forelimb. Brown, (1906) described articulated sternals that are attached along the middle of both. Johnson and Ostrom attach these bones just at the upper tips with a gap on the caudal end. The coracoids don't really attach to the sternals. The lower tip of the coracoid touches the upper outer edge of the sternal. This is very important since it determines the width of the pectoral girdle. The pectoral girdles, as stated above, have an inherent 'U' shape and do not flare outward.

Greg Paul draws a cartilage that rests over the sternals. This would act as a sliding joint for the coracoid to pass along. Ken Carpenter (pers. comm.) has dissected lizards and crocodilians and has seen a cartilage rim on the medial side of the coracoids. The rim is slotted cartilage (gliding joint) against the cartilage bone that would fit above the sternals. The cartilage is of the same type as that of a chicken drumstick. Both sides of this gliding joint are bound by fibrocartilage. This fibrocartilage allows some movement along the gliding joint, but not, in my opinion, to the degree stipulated by Greg Paul.

The scapula is oriented nearly parallel to the vertebral column. Rib reconstruction is determined by the width of the pectoral girdle, and not the other way around. A narrow chest results, not Johnson's and Ostrom's barrel chest.

What difference is there between a wide or narrow width of the pectoral girdle? (figure 6, 7). The most important factor is the glenoid. Because the 'U' shaped unit of the pectoral girdle is forward of the first rib, the glenoid allowed the humerus to move freely from the rib cage. If the scapula were held in a wide gauge, the coracoids would be too far apart, would be separated from the sternals and the glenoid would have faced too far laterally and not enough ventrally. This would force the humerus into a horizontal sprawl. But when the pectoral girdle is narrow, with the coracoids close to each other and attaching to the sternals, the glenoid faces lower. This inhibits a "sprawling" humerus.

What part of the humeral head has the most bone, or which plane does it most favor? The humeral head is not oval, but has a more rectangular shape. The largest part of the head supports a vertical movement, not a horizontal movement, i.e. the bone mass for a rotational horizontal movement is lacking but the bone mass supports a 'vertical' movement. Since the humeral head would most likely have had a large cartilage 'head' along with a cartilage 'socket' in the glenoid, similar to that of a chicken wing, it would have to have mimicked the surrounding bone for the most part.

If the forelimb had a sprawl (Johnson and Ostrom suggest the ulna and radius would have been positioned at a 45° angle to each other), then the humerus would have been held nearly horizontal (or as Johnson and Ostrom have the humerus with a slight downward curve). Looking from above, we can see an arch that the humerus would have taken while moving forward to aft. Because the ulna has a large olecranon, it would have prevented the ulna from being completely vertical. The forelimb would have been held with a bend in it. For the humerus to be held horizontally, the ulna and radius condyles would have had to have been oriented ventrally on the humerus, similar to that of the femur on birds (in birds the femur is held, for the most part, horizontally, with the lower leg doing the most visible movement). Also the condyles would have had to have created a "center" for the ulna and radius to cycle around the condyles while the humerus was moved forward and aft.

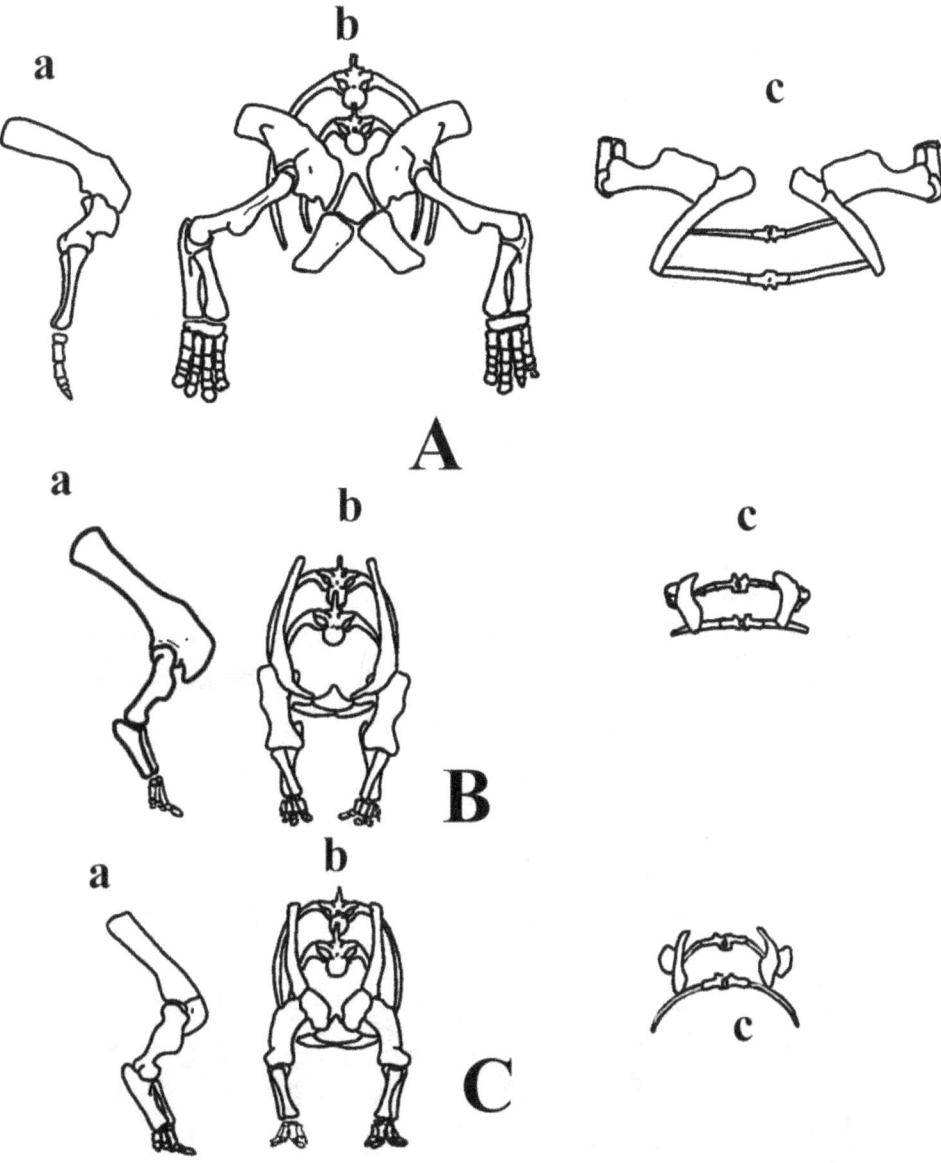

Figure 6). Side view of *Torosaurus* after Johnson and Ostrom (A) in (a) lateral view, (b) front view and (c) top view; B) after Bakker in (a) lateral view, (b), front view and (c) top view; C), after Paul in (a) lateral, (b) front, and (c) top view.

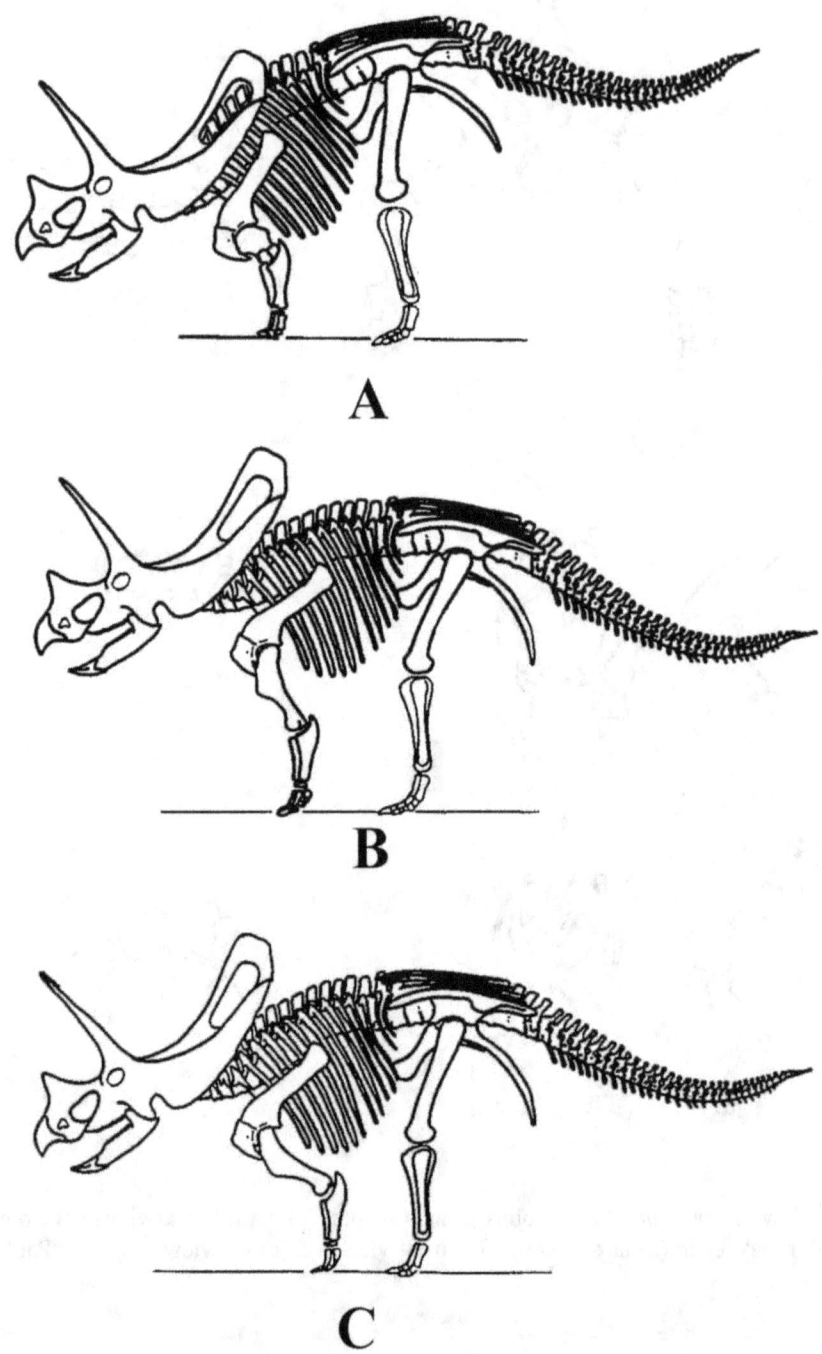

Figure 7). *Torosaurus* skeleton showing stance after (A) Johnson and Ostrom, note how close the front end of the skeleton is to the ground; (B) Bakker, and (C) Paul.

The foot is the only part of the forelimb that didn't move when it was on the ground. Walking through a footstep we can see what the leg does while in stride. The foot stays stationary; that much we know for sure by looking at the manus prints in *Tetrapodosaurus* which has no smearing of the toes. The foot was held tilted slightly outward and just a little outside the pes. The movement of the body during the stride (with a stationary foot) determines how the leg moves and not the movement of the leg itself. In using Johnson and Ostrom's model, during the beginning of the stride the ulna and radius is slightly outside the outer edge of the humerus, in the middle of the stride, the ulna and radius would be inside the humerus. At the end of the stride, the ulna radius is again outside the humerus.

During this phase the ulna and radius move around the humerus condyles, and not just fore and aft as if walking with an upright posture. Instead, they move the humerus in a horizontal arch, with the ulna and radius vertical to the ground and the foot swinging freely with the lower leg and foot. However, the foot must be considered stationary, and the pectoral girdle must be moved forward and backward to accurately appreciate ceratopian front limb movement.

ALTERNATIVES

The reason for the sprawling stance, according to Johnson and Ostrom, is to compensate for the heavy skull. If the skull was 1/3 the weight of the animal, then at one point of the stride one foot would have to be off the ground and the animal would have had to do a one-handed-pushup while walking and the amount of muscle tension and strain would have been considerable.

If the leg was held nearly vertical, the stance Bakker advocates for ceratopians, the humerus lies in a straight line with the pes, with the 'elbow' tilting inward against the body, the leg not bowing at all. The chest is more barrel-shaped and the feet lie inside the midline of the body. Bakker's ceratopian leg moved in the same fashion as a rhino.

Using the stance that Greg Paul favors for ceratopians, the humerus would have a slight outward bend, the humeral head would fit perfectly into the glenoid and the humerus would move fore and aft in an inverted arch. The extent of flexure of the humerus during the walking phase would be nearly vertical and then the end phase would be nearly at a 45 ° angle. The inside of the humeral head would still be held inside the body. The leg would still have a bend due to the large head on the ulna, and the ulna and radius would cross slightly. As the leg was moved, the scapula-coracoid would move slightly along the cartilage bone on top of the sternals synchronously with the fore and aft movement of the leg.

CONCLUSION

So, which stance works best? In my opinion, to answer this question, first ask, "Which stance corresponds with trackways of *Tetrapodosaurus*?" (figure 8). A sprawling stance, as used by Johnson and Ostrom (*Torosaurus* mount), is too wide. To compensate for the fore foot print, the ulna and radius must match a *Tetrapodosaurus* trackway, but the in-between stride would make it impossible. The upright stance, initial stride and end portion of the stride are all consistent with the nature of *Tetrapodosaurus* prints. The Bakkerian stance would have required a chest that was too wide proportionately to compensate for the prints, and the legs would have had to have been bowed out. The stance that works best is the one that Greg Paul supports. The legs bow out a little then come inside to match the tracks of *Tetrapodosaurus*.

Acknowledgements

I would like to thank Ken Carpenter, Darren Tanke, Larry Barnes and George Olshevsky whom I've had several talks about the ceratopian stance.

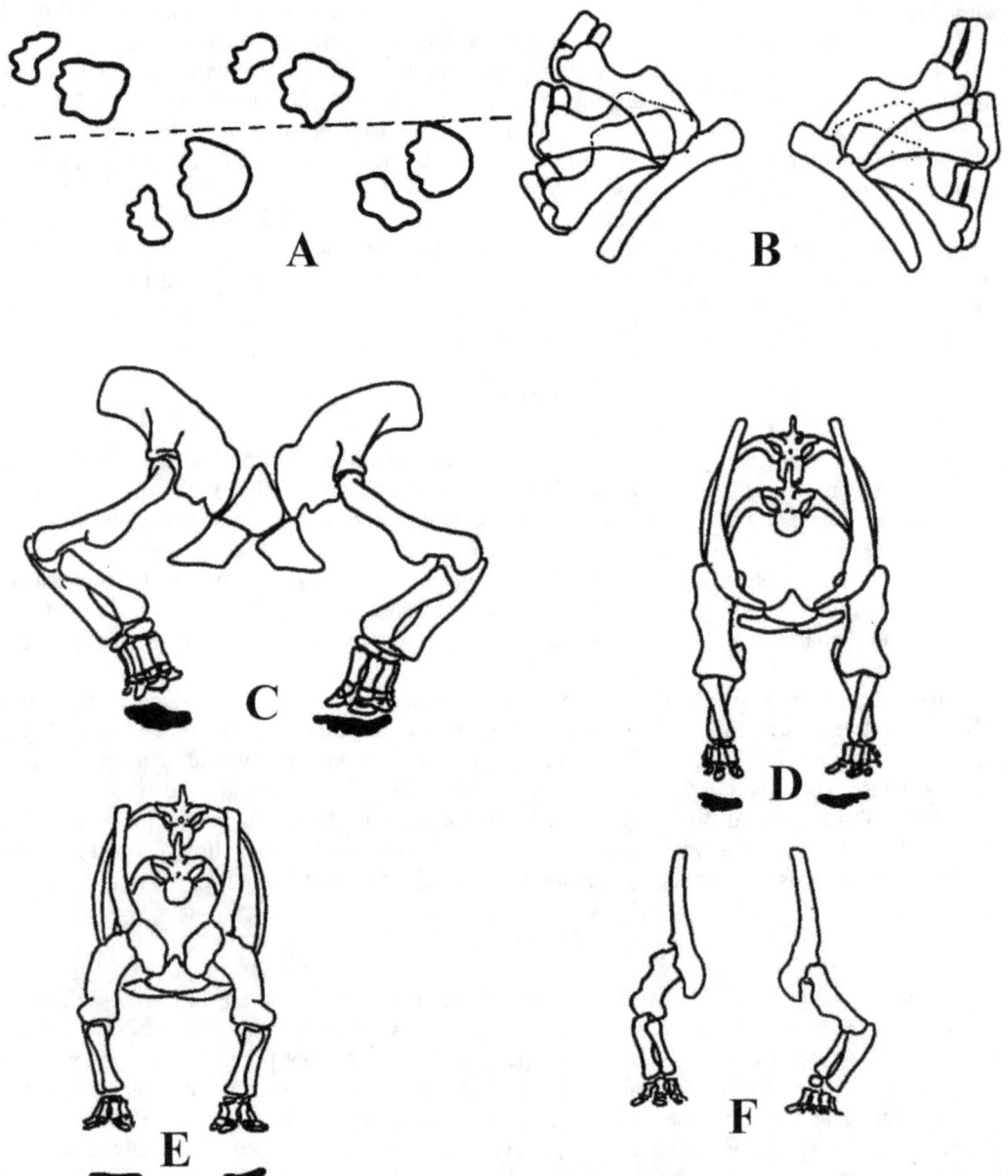

Figure 8) A). Top view of the footprints of *Tetrapodosaurus*. B) Modifying the stance of Johnson and Ostrom to fit the prints in top view. The right side shows how the ulna and radius would move in a natural phase, and the left side shows how the ulna and radius would move if they were not floating around the condyles of the humerus. C) Front view modified after Johnson and Ostrom to fit the trackway. D). Stance of Bakker theory showing the position of the front legs would have to be moved outward. E). Paul showing no modification is needed to fit the tracks. F) Incorrect mounted skeleton of *Triceratops prorsus*, (BROWN, 1906).

Figure 9). Life restoration of *Torosaurus* in front and lateral view.

71

Bibliography:

Bakker, R. T., 1986, The Dinosaur Heresies, New Theories Unlocking the Mystery of the Dinosaurs and their extinction. William Morrow and Company, Inc. New York: 481pp.

Brown, B. B., 1906, New notes on the ostelogy of *Triceratops*: Bulletin of the American Museum of Natural History, v. 22, p. 297-300.

Brown, B. B., and Schlaikjer E. M., 1940, The structure and relationships of *Protoceratops*: Annual of the New York Academy of Science, v. 11, article 3, p. 133-266.

Carpenter, K., 1989, Common mistakes in the mounting of fossil skeletons: Journal of Vertebrate Paleontology, v. 9, supplement to n. 3, Abstracts of papers, Forty-ninth Annual Meeting, Society of Vertebrate Paleontology, Vertebrate Paleoontology Laboratory, The University of Texas, Austin Texas, November 2-4, p. 15A.

Carpenter, K., Madsen, J. H. jr., and Lewis, A., 1994, Mounting of fossil vertebrate skeletons: In: Vertebrate Paleontological Techniques, edited by Leiggi P., and May P., Volume One, Chapter 11, p. 285-322.

Dodson, P., 1996. The Horned Dinosaurs. Princeton University Press: 346pp.

Hatcher, J. B., Marsh O. C., and Lull R. S., 1907, The Ceratopsia: Monographs of the United States Geological Survey, v. 49, p. 1-300.

Hunt, A. P., Lucas, S. G., and Lockley, M. G., 1997, On trackways from the Early Permian of New Mexico: New Mexico Geology, v. 19, n. 2, p. 54.

Johnson, R. E., and Ostrom, J. H., 1990, Biomechanical analysis of forelimb posture and gait in *Torosaurus*: Journal of Vertebrate Paleontology, v. 10, supplement to n. 3, Abstracts of papers, Fifieth Annual Meeting, Society of Vertebrate Paleontology, Museum of Natural History, University of Kansas, Lawrence, Kansas, p. 29A-30A.

Johnson, R. E., and Ostrom, J. H., 1991, A refined model of *Torosaurus* forelimb posture and gait with implications for the Ceratopsia: Journal of Vertebrate Paleontology, v. 11, supplement to n. 3, Abstracts of Papers, Fifty-First Annual Meeting, Society of Vertebrate Paleontology, San Diego State University and San Diego Natural History Museum, San Diego, California, October 24-26, p. 39A.

Johnson, R. E., and Ostrom, J. H., 1995, The forelimb of *Torosaurus* and an analysis of the posture and gait of ceratopsian dinosaurs: In: Functional Morphology in Vertebrate Paleontology, edited by Thomason, J. J., Cambridge University Press: 205-218.

Lockley, M., 1991. Tracking Dinosaurs. A new look at an ancient world. Cambridge University Press: 238pp.

Lull, R. S., 1908, The cranial musculature and the origin of the frill in the ceratopsian dinosaurs American Journal of Science, 4th series, v. 25, p. 387-399.

Lull, R. S., 1933, A revision of the Ceratopsia or Horned Dinosaurs: Memoirs of the Peabody Museum of Natural History, v. 3, part 3, p. 1-175.

Lull, R. S., 1934, Skull of *Triceratops flabellatus* recently mounted at Yale: The American Journal of Science, 5th series, v. 28, p. 439-442.

Molnar, R. E., 1973, The Cranial Morphology and mechanics of *Tyrannosaurus rex* (Reptilia; Saurischia). Master Thesis: University Microfilms International, 73-18,639, 451pp.

Ostrom, J. H., 1966, Functional morphology and evolution of the Ceratopsian Dinosaurs: Evolution, v. 20, p. 290-308.

Ostrom, J. H., and Wellnhoffer, P., 1986, The Munich specimen of *Triceratops* with a revision of the genus: Zitteliana, v. 14, p. 111-158.

Paul, G. S., 1987, The Science and art of restoring the life appearance of Dinosaurs and their relatives: In: Dinosaurs Past and Present, v. II, p. 4-49.

Paul, G. S., 1991, Giant horned dinosaurs did have fully erect forelimbs: Journal of Vertebrate Paleontology, v. 11, supplement to n. 3. Abstracts of Papers, Fifty-First Annual Meeting, Society of Vertebrate Paleontology, San Diego State University and San Diego Natural History Museum, San Diego, California, October 24-26, p. 50A.

Russell, L. S., 1935, Musculature and function in the Ceratopsia: Bulletin of the National Museum of Canada, v. 77, p. 39-44.

For the Dinosaur Collector and Enthusiast

PREHISTORIC TIMES

April/May 1999   No. 35

Jurassic Park:
The Ride

THE PT
INTERVIEW:
Gregory S. Paul

Also:
Donald F. Glut
D.W. Miller

Ford, T. L., 1998, How to Draw Dinosaurs. Those hard-headed boneheads, the Pachycephalosaurs: Prehistoric Times, n. 31, p. 12-15.

Chapter 11

Those hard-headed boneheads, the Pachycephalosaurs

Pachycephalosaurians are a very strange breed of dinosaurs to say the least. Over-aggressive males bashing each other's brains with their large round-thick heads making sounds akin to rams dueling for territory is the picture usually portrayed by artists. The affections of a female echoing through out a valley, whilst a group of females watch timidly comes to mind. A placed scene indeed, but is that a correct one? Could they really butt heads? Also, pachycephalosaurs are often depicted with the wrong body shape, one typical of an Ornithopod. Actually, they had a wide body. This will be a two part article. The first part will address the domes, the second; head butting.

Part One: The Domes...

There are two families in the pachycephalosauria, the flat-headed Homalocephalidae and the dome-headed Pachycephalosauridae. Both have the same general body shape, but with very different skulls. The Homalocephalians are *Homalocephale*, the namesake of the family (figure 1a), *Goyocephale*, *Micropachycephalosaurus* (The dinosaur with the longest name), and *Wannanosaurus*. They do have little 'knobs' at the back of the skull. There is a premaxillargy fang and a diastema (groove) behind the fang followed by the maxillary teeth. The dentary also has a pair of fangs. *Homalocephale* lacks the front end of the skull, but *Goyocephale* is more complete, thus the whole skull of homalocephalians is known. **(Editors note: The validity of this family has been challenged by several paleontologist and is no longer considered valid).**

The 'domed' dome heads have domes that differ from genus and species as well as genus to genus. They have different apexes for the dome. The dome shape itself varies from genus to genus, as well as the size of knobs; also whether there is a 'shelf' at the rear of the skull. *Gravitholus* is known from just a dome (figure 1b). The dome lacks knobs (thou this may be due to its fragmentary nature), lacks a shelf and has a dome that has its apex near the rear of the skull. *Pachycephalosaurus*, (figure 1c) for which the family is named after, has a long snout (with the premaxilla missing, so it is not known if it had a premaxillary fang or diastema), rounded 'knobs' on the snout and rear end of the skull. It lacks a shelf, and has the apex of the dome well behind the orbits. *Prenocephale*, (figure 1d) has a short complete skull, lacks a shelf, and has small knobs on the skull. The apex of the dome is at or just behind the orbits. When viewed from above, the snout is very slender, it has a small premaxillary fang and a diastema. An interesting side note is that when the skull was found it had the dome facing outward from a small sandy hill. The skull was pulled, intact, out of the hill. *Stegoceras*, (figure 1e), has a short snout, little knobs and a shelf. The apex is just behind the orbit. There are 3 small premaxillary teeth, none being a fang, and a very small diastema. *Stegoceras* is the most abundant pachycephalosaur. Its fragmentary domes have been found by the dozens. The domes vary in both size and shape from specimen to specimen, indicating growth stages and sexual dimorphism. *Tyocephale* (figure 1f) is known from a partial skull that lacks the snout. The dome lacks knobs, a shelf, and is very narrow when viewed from above.

The history of *Stygimoloch* is like a Sherlock Holmes novel. What was found first was a spiky squamosal, then just a thin dome, then a specimen with both. New evidence from several different specimens shows just how the skull appeared. *Stygimoloch* has the most elaborate skull of pachycephalosaurians. The skull is long and similar to *Pachycephalosaurus* but has spiky, not rounded knobs. The largest spikes are on the posterior edge of the dome. The skull lacks a shelf and has a short, thin dome. The apex is hard to place, but it seems to be behind the orbit. *Stygimoloch spinifer* was originally described by Galton and Sues (1983) from a partial left spiked squamosal that bordered the posterior end of the skull (UCMP 119433, figure 2a), another squamosal (YPM 335) was originally attributed to *Triceratops* by Marsh. Giffen, et al (1987), described *Stenotholus kohleri* (MPM 711, figure 2b), from a diagnostic thin dome. As the paper describing *Stenotholus* was in press, the Milwaukee Museum field crews discovered another dome of *Stenotholus* with the spikes of *Stygimoloch* attached to it (MPM 811, figure 2c) and Gabriel and Berghaus (1988) sunk *Stenotholus* into *Stygimoloch*. Mark Goodwin, Emily

Buchholz and Rolf Johnson have recently described more *Stygimoloch* specimens as well as MPM 8111 in more detail. Mike Triebold (a private collector) has found a partial skull and skeleton of *Stygimoloch*. This

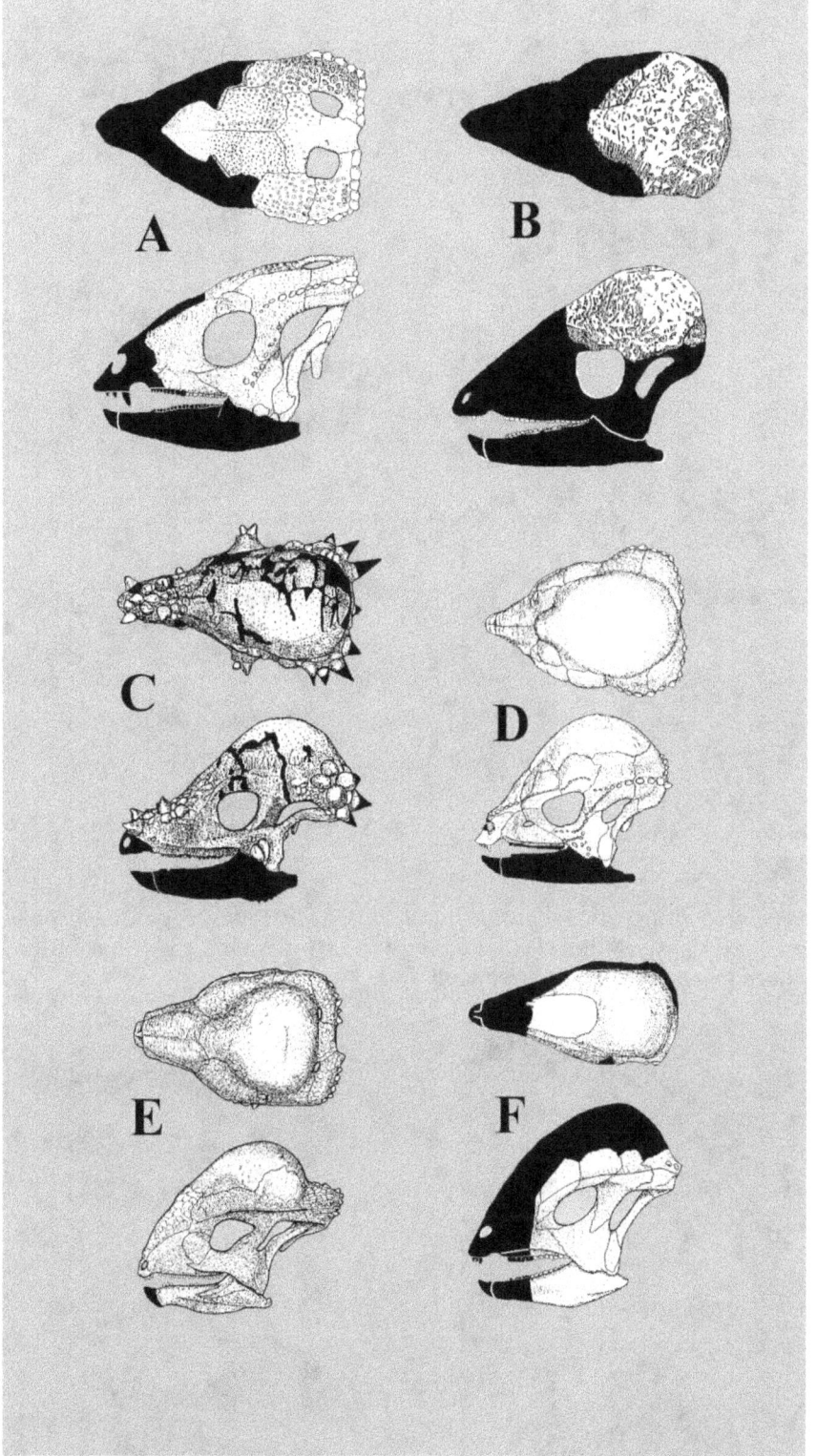

Figure 1), Domes of pachycephalosaurs, dorsal and side view; A) *Homalocephale*; B) *Gravitholus*; C) *Pachycephalosaurus*; D) *Prenocephale*; E) *Stegoceras*; and F) *Tyocephale*.

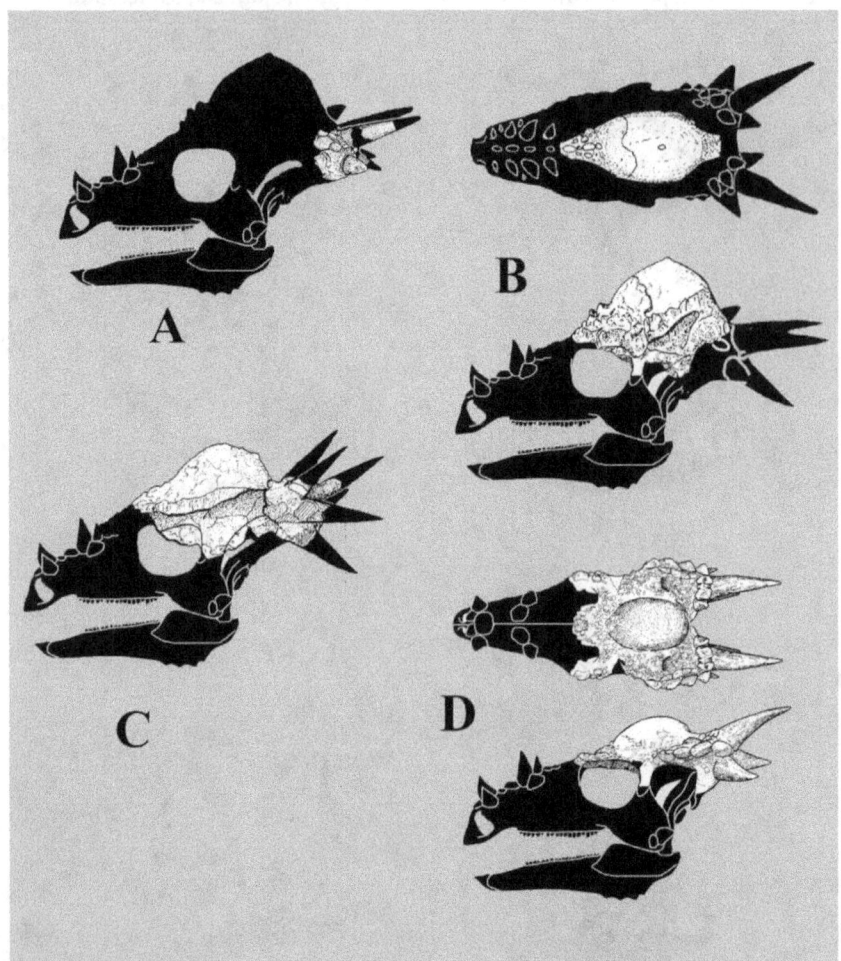

Figure 2), *Stygimoloch* skulls; A) *Stygimoloch spinifer*, UCMP 119433; B) *Stenotholus kohleri*, MPM 7111; C) *Stygimoloch spinifer* MPM 8111: (**Editors note: D) is from a cast I have from Fort Peck. I used this specimen to modify the first three skulls**).

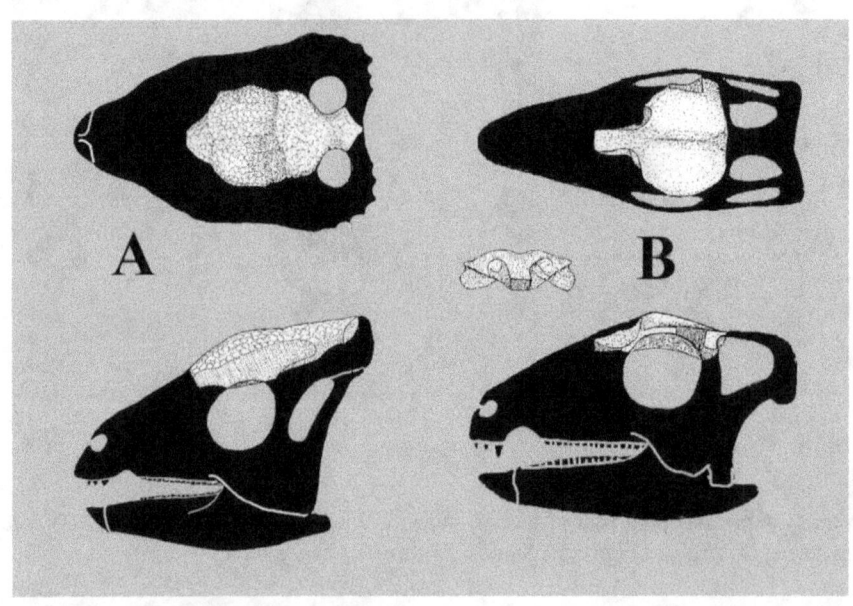

Figure 3), A) *Ornithotholus* skull in top and side view (after Galton & Sues, 1983); B) *Yaverlandia* in top, rear, and side views (after Galton, 1971).

specimen is the most complete large pachycephalosaur known. The skull lacked the dome, which is unusual, the premaxilla, and tip of dentary. The squamosals were in place and with MPM 8111 shows just how the spikes were orientated.

Originally it was displayed with the spikes facing upward, but the newer specimens show that the spikes were horizontal. The lower jaw was found and the posterior end has small spikes, unlike the smooth lower jaw of *Stegoceras* and *Goyocephale*. Mike made a cast of the skull with a dome of *Pachycephalosaurus*. I've talked to him about his interpretation and at the time he hadn't heard of *Stygimoloch*, so I can't really fault him for putting on the wrong dome. From the Sandy Site, where the new pachycephalosaur came from, 3 types of domes were found, a small dome, a dome similar to *Pachycephalosaurus* and one thin dome, which probably belongs to *Stygimoloch*, so he had three ways to go, and he decided to go with the *Pachycephalosaurus* dome. Since this specimen hasn't been described yet, I've refrained from doing an illustration of it. It should be noted that any 'spiked' pachycephalosaur is *Stygimoloch* and should have a thin done, not large round dome of *Pachycephalosaurus* like the one at the Philadelphia Academy of Science.

Because the snout is missing it isn't known if there was a fang or diastema. Some believe that *Stygimoloch* and *Pachycephalosaurus* are one and the same, but *Pachycephalosaurus* are one and the same, but *Pachycephalosaurus* has a long skull with the dome well over the occipital condyle, and although *Stygimoloch* also has a long skull, the dome shape and the position of the squamosal do not support the two belonging to the same genus. (**Editors note, I know believe the pachycephalosaur that Mike Triebold had is not a *Stygimoloch*, but is a *Pachycephalosaurus* with spikes, instead of knobs. I also now believe that the knobs of *Pachycephalosaurus* are worn down and should be spikes. This article was written long before *Dracorex* and the theory of *Dracorex*, *Stygimoloch* and *Pachycephalosaurus* being a growth series, which I don't believe.)**

Ornatotholus (figure 3a) has a flat head for a pachycephalosaurid pachycephalosaur. It is known from a very low dome and is very small. Some believe this to be from a juvenile *Stegoceras*, but others believe it to be a valid genus. It looks as if it might have had a shelf but lacked knobs. *Yaverlandia* (figure 3b) is also a flat-headed Pachycephalosaurid pachycephalosaur. It is a very very low bifurcated dome. I have hard that someone has questioned the family placement in the Pachycephalosauridae for *Yaverlandia*, but that is all I know as of this writing. (**Editors note: It has since been shown that *Yaverlandia* is a Troodontid theropod [Naish, 2011])**,

The body of pachycephalosaurs differs from 'normal' ornithopods (I don't believe that pachycephalosaurs are ornithopods, but are instead closer related to the heteradontosaurs and the ceratopia, but that is a different subject), instead of a thin body, they have a very wide body (figure 4). The skull is at nearly a right angle to its stocky neck. When viewed from above, the front of the 'body' is narrow, then quickly widens, then narrows a bit towards the pelvis. The pelvis is wider at the posterior end than at the anterior. The first caudal vertebrae are widest then quickly narrows to the tail. The tail is very thin when viewed from above. From the side the body is short in height, with the tail being moderate in size. The tail also has basket shaped tendons that stiffen the tail. No complete skeleton of a pachycephalosaurs is known, but *Homalocephale*, *Stegoceras* and Triebold's *Stygimoloch* are the best known skeletons. The front legs are short, but unfortunately the hands are not known. The feet have 4 metatarsals, with first being vey short.

It has recently been shown that the Madagascar pachycephalosaur; *Majungatholus* actually belongs to an Abelisaurid theropord (Samson et. al., 1998).

Part Two: Butt me no Butts…

I haven't believed that pachycephalosaurs butted their heads for years now. In a conversation with Ken Carpenter some time ago we discussed this possible activity. We both agreed (for both the same and different reasons) that they didn't. Mark Goodwin, in a Paleoworld TV episode also questioned if they butted their head. I wanted to do an article on this years ago, but Ken told me that he had a paper in press. I decided he was the better one to write it and left it as that. His paper was published late last year and Goodwin et al was just published in the Journal of Vertebrate Paleontology, volume 18, number 2. I will be using both Carpenter's and Goodwin's beliefs as to why pachycephalosaurs did not butt heads as well as making comments on my own.

A quick history of the butting theory first. Ned Colbert first surmised head-butting in his 1955 book, Evolution of the Vertebrates. Colbert states…Perhaps (as a very wild surmise) the skull was used as a

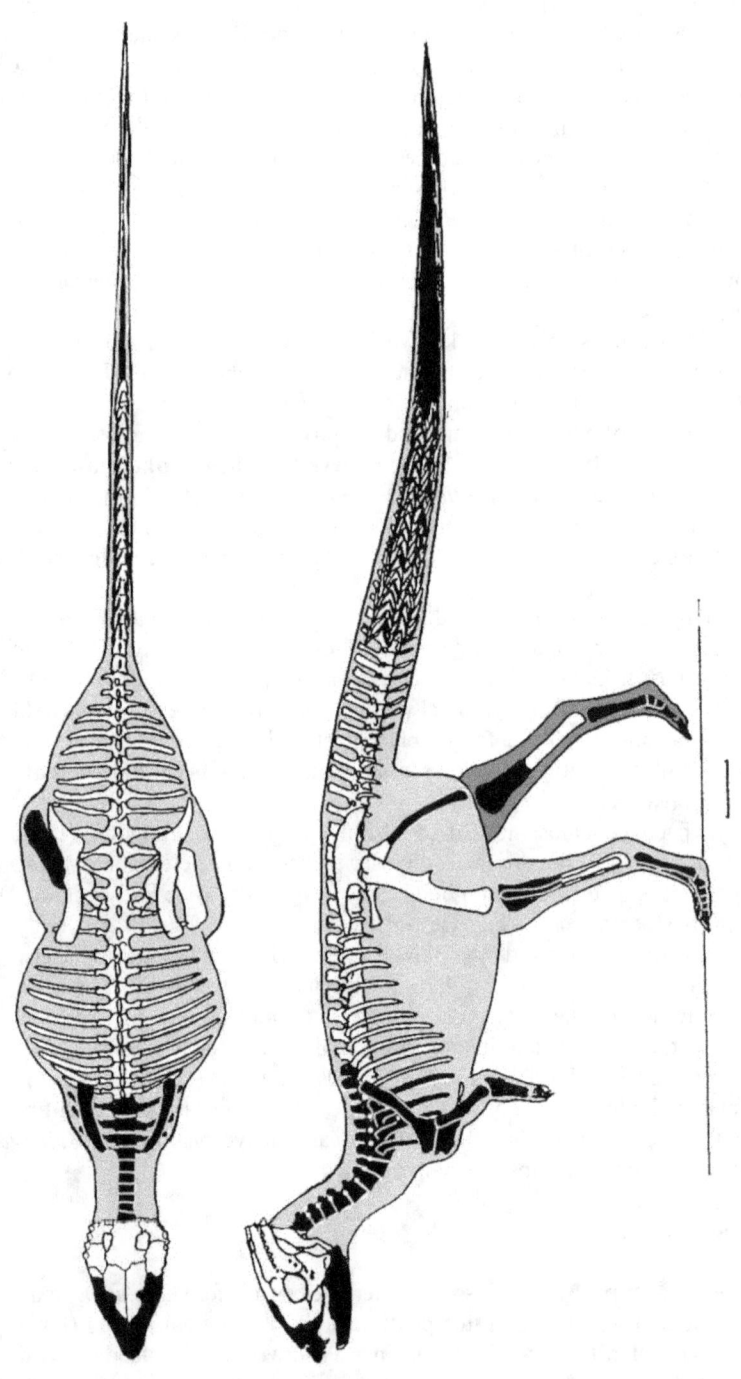

Figure 4), A) Top view of *Homalocephale*; B) side view of *Homalocephale*; Both after Maryansak and Osmolska, 1974, and Greg Paul.

sort of battering ram... Galton wrote longer papers favoring the battering ram hypothesis (1970, 1971) and Sues (1978) wrote an extensive study about the skulls of *Stegoceras* and concluded that they could have practiced head butting, but the limited surface area and lack of any self-correcting mechanism for misaligned heads made it difficult and suggested flank butting. The skull was thought to be a solid dome and the braincase was well ossified allowing it to take a good blow to the head. Others have also commented that the skeleton would have been held totally horizontal which would have absorbed the shock of blows (figure 5c).

Pachycephalosaurs are thought to have butted their heads like a big horn sheep (figure 6a). Big horn sheep don't butt heads directly, but butt two horns. The horns are flat at the top which supports a larger surface area for the horns to butt. Bison also butt heads, but they have a thick tuff of hair to cushion the blows. Pachycephalosaurs, as noted in part one, have different apexes for the dome and different dome shapes (figure 6b-g). They all have just one surface area for the butting to take place, not two as in Big Horn Sheep. The Sheep can line up the attack by looking forward. Pachycephalosaurs don't have stereo vision so it's hard to look forward, and they can not keep an eye on each other as the head is lowered because the dome over hangs the orbits. The domes, according to Goodwin et al., is not solid but porous. This is an indication of fast growth and could not withstand head butting. It would have been very difficult for them to butt heads because they could not keep an eye on the smaller surface area and keep their combatant opponent lined up. Even head 'pushing' would not have been easy due to the difficulty of lining up the heads.

Picture yourself playing pool. You roll the cue ball to knock the eight ball into a pocket via a straight line. Now, if both you and your friend roll two balls toward each other trying to hit dead on, you'll see how difficult it really is. The balls will sometimes head dead on, but most of the time will glance off one another. This is what would happen if pachycephalosaurs butted their heads. Also big horn sheep hit via their horns, which are covered by a horny sheath. Pachycephalosaurs don't have a horny sheath to protect their heads, they'd first hit skin to skin and then would glance off one another' s knobs, cutting themselves. Something like falling down and skinning your knee on gravel.

Ken Carpenter also portrayed the skeleton as it really was. The neck has a natural 'S' shape and the body has a curved back. If the head was lowered, the neck couldn't have been held straight. If it was, the neck vertebrae would have been displaced. The back would have been lower and could not have been in a straight line as often depicted. The vertebral column could not withstand a butting attack (figure 5a).

What then were the domes used for? Sexual dimorphism? Heat regulation like Keith Rigby Jr surmises? Or flank butting? Both Ken Carpenter and Goodwin et al. come to this conclusion. Ken elaborates that the animals would have stood parallel to one another, either facing each other or in the same direction, and butted each other in the sides (Figure 7). The wide body would protect the vital internal organs. *Stygimoloch* plays a big part in both papers. Another thing to take into account is that the eyes are on the sides of their heads, and attacking parallel to one another would have been very easy, since they could 'see' each other without turning their head much. Ken believes the horns were held slightly upward and could have been used in the flank butting, maximizing the pain. But Goodwin et al show the horns were horizontal and could not have been used in flank butting. They believe that *Stygimoloch* simply lowered its head to show its horns in an agnostic display.

Both Carpenter and Goodwin et al. papers say the Homalocephalids may have butted heads, but due to the limbed skull depth and protection this is also unlikely.

Ken Carpenter writes in his conclusion...Pachycephalosaurids probably began with much display, possibly with head bobbing and presenting their head in lateral profile to emphasize their side of the dome or horns. This type of threat display, signals a willingness to fight, and draws attention to the offensive weapon, in this case dome and horns

By standing parallel or reverse parallel, each animal may have tried to intimidate the other by signaling a "determination to fight". Eventually, one or the other pachycephalosaurid may have directed a blow or blows to the side of the other. Further threat and intimidation display punctuated by blows may have followed, until or the other signaled submission and appeasement...I couldn't have put a better scenario together myself.

Head butting in pachycephalosaurs should be put to rest and flank butting should be the correct interpretation.

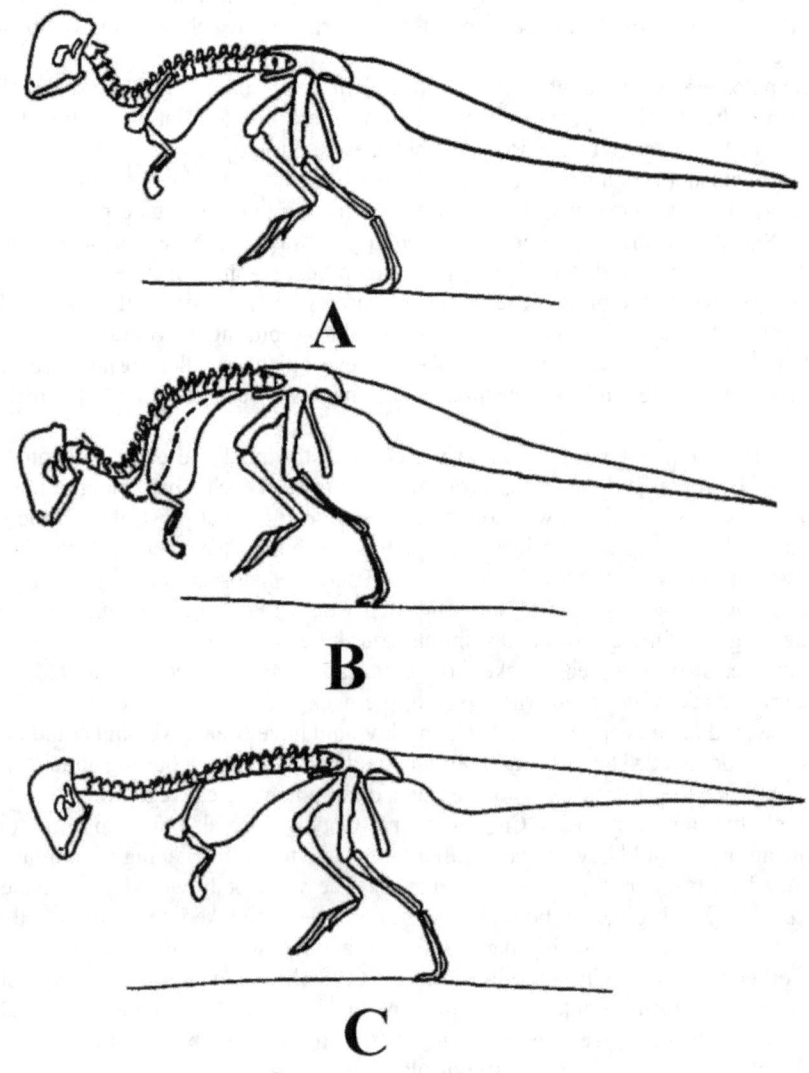

Figure 5), Side view of the skeleton of *Stegoceras* in normal stance; B) side view of skeleton showing how low the body has to be if it head butted, the black lines show how far the vertebral column would go if the head butted (Both after Carpenter, 1997); C) incorrect drawing of *Stegoceras* with a horizontal vertebral column.

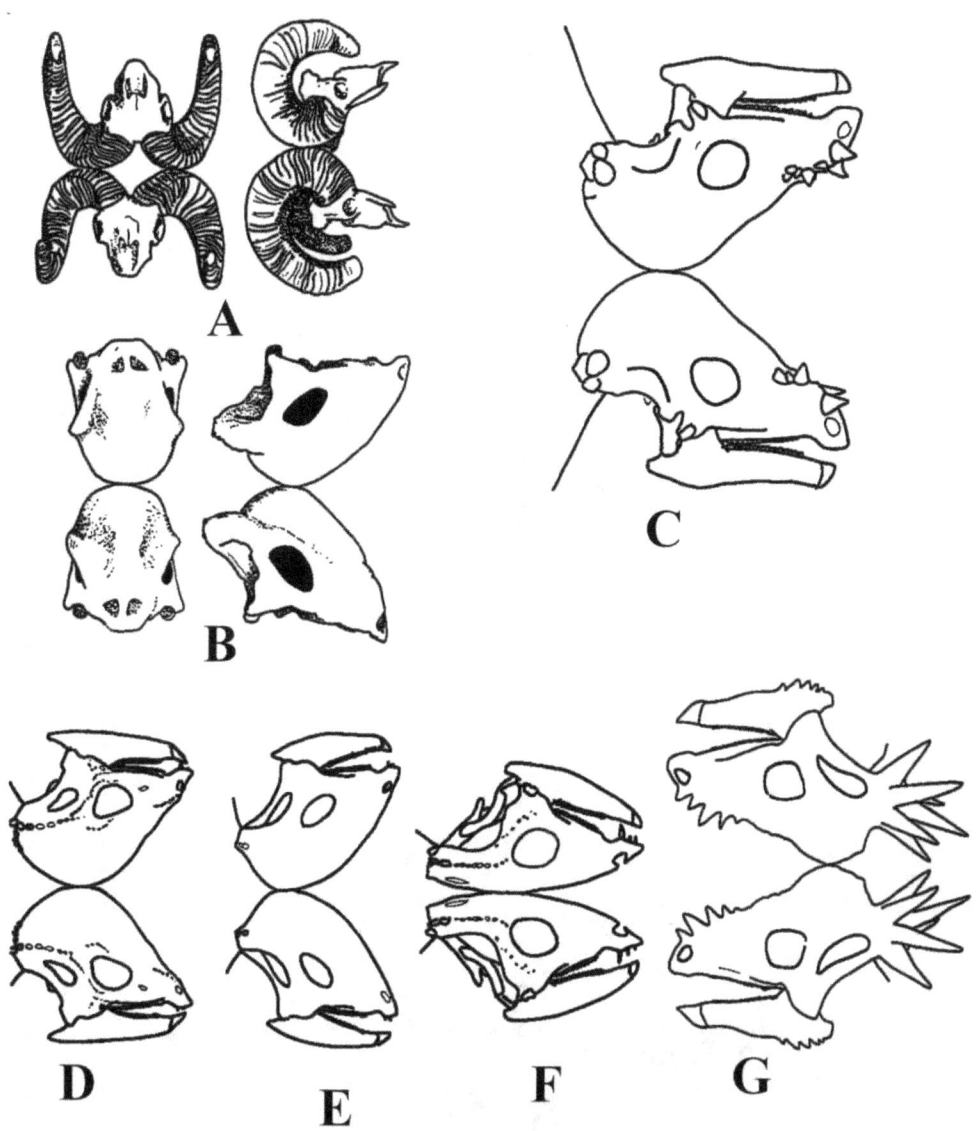

Figure 6), Front and side views of skulls; A) Big Horn Sheep (after Carpenter, 1997); B) *Stegoceras* showing the 'butting' surface (after Carpenter, 1997), C) *Pachycephalosaurus*; D) *Prenocephale*; E) *Tylocephale*, F) *Homalocephale*, and G) *Stygimoloch*.

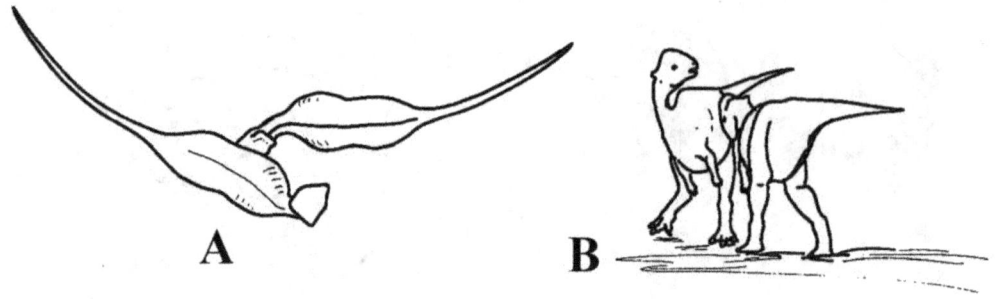

Figure 7), A) Top view of *Steogceras* flank butting; B) front view of flank butting (Modified from Carpenter, 1997).

Figure 8), Life restoration of *Homalocephale*.

Bibliography

Brown, B. B., and Schlaikjer, E. M., 1943, A study of the Troodont Dinosaurs with the description of a new genus and four new species: Bulletin of the American Museum of Natural History, v. 82, article 5, p. 120-149.

Carpenter, K., 1997, Agonistic behavior in pachycephalosaurs (Ornithischia: Dinosauria): a new look at head-butting behavior: Contributions to Geology, University of Wyoming, v. 32, n. 1, p. 19-25.

Gabriel, D. L., and Berghaus, C. B., 1988, Three new specimens of *Stygimoloch spinifer* (Ornithischia: Pachycephalosauridae) and behavior inferences based on cranial morphology: International Symposium on Vertebrate Behavior as Derived from the Fossil Record, September 8-10, 1988. Museum of the Rockies, Montana State University, Bozeman, MT, 59717, unnumbered.

Galton, P. M., 1970, Pachycephalosaurids, dinosaurian battering rams: Discovery, v. 6, n. 1, p. 23-32.

Galton, P. M., 1973, Redescription of the skull and mandible of *Parksosaurus* from the Late Cretaceous with comments on the family Hypsilophodontidae (Ornithischia): Life Sciences Contribution Royal Ontario Museum, v. 89, p. 1-21.

Galton, P. M., and Sues, H.-D., 1983, New data on pachycephalosaurid dinosaurs (Reptilia: Ornithischia) from North America: Canadian Journal of Earth Sciences, v. 20, p. 462-472.

Giffin, E. B., Gabriel, D. L., and Johnson, R. E., 1987, A new pachycephalosaurid skull (Ornithischia) from the Cretaceous Hell Creek Formation of Montana: Journal of Vertebrate Palaeontology, v. 7, n. 4, 398-407.

Gilmore, C. W., 1924, On *Troodon validus*, An Ornithopodous Dinosaur from the Belly River Cretaceous of Alberta, Canada: Bulletin of the University of Alberta, Department of Geology, v. 1, p. 1-43.

Goodwin, M. B., Buchholtz, E. A., and Johnson, R. E., 1998, Cranial anatomy and diagnosis of *Stygimoloch spinifer* (Ornithischia: Pachycephalosauria) with comments on cranial display structures in agonistic behavior: Journal of Vertebrate Paleontology, v. 18, n. 2, p. 363-375.

Goodwin, M. B., and Johnson, R. E., 1995, A new skull of the pachycephalosaur *Stygimoloch* cast doubt on head butting behavior: Journal of Vertebrate Paleontology, v. 15, supplement to n. 3, Abstracts of Papers, Fifty-Fifth Annual Meeting Society of Vertebrate Paleontology, Carnegie Museum of Natural History, Pittsburgh, Pennsylvania, November 1-4, p. 32a.

Maryanaska, T., and Osmolska, H., 1974, Pachycephalosauria, a new suborder of Ornithischian Dinosaurs: Palaeontologica Polonica, v. 30, p. 45-102.

Olshevsky, G., 1988, *Stenotholus* is *Stygimoloch*: Archosaurian Articulations, v. 1, n. 4, p. 28-29.

Sampson, S. D., Krause, D. W., Forster, C. A., and Dodson, P., 1996, Non-Avian theropod dinosaurs from the Late Cretaceous of Madagascar and their paleobiogeographic implications: Journal of Vertebrate Paleontology, v. 16, supplement to n. 3, Abstracts of Papers, p. 62A.

Sampson, S. D., and Witmer, L. M., 1999, Novel narial anatomy in ceratopsid dinosaurs. Journal of Vertebrate Paleontology, v. 19, supplement to n. 3, Abstracts of Papers, Fifty-ninth annual meeting, Society of Vertebrate Paleontology, Adams Mark Hotel, Denver, Colorado, October 20-23, p. 72A.

Sampson, Scott D., Witmer, L. M., Forster, C. A., Krause, D. W., O'Conner, P. M., Dodson, P., and Ravoay, F., 1998, Predatory Dinosaur remains from Madagascar: Implications for the Cretaceous Biogeography of Gondwana: Science, v. 280, p. 1048-1051.

Sues, H.-D., 1980, A pachycephalosaurid dinosaur from the Upper Cretaceous of Madagascar and its paleobiogeographical implications: Journal of Paleontology, v. 54, n. 5, p. 954-962.

Triebold, M., 1997, The Sandy Site: Small Dinosaurs from the Hell Creek Formation of South Dakota: In: Dinofest International, Proceedings of a Symposium sponsored by Arizona State University. A Publication of The Academy of Natural Sciences, edited by Wolberg D. L., Sump E., and Rosenberg G. D., p. 245-248.

Triebold, M., and Russell, D. A., 1995, A new small dinosaur locality in the Hell Creek Formation: Journal of Vertebrate Paleontology, v. 15, supplement to n. 3, Abstracts of Papers, Fifty-Fifth Annual Meeting Society of Vertebrate Paleontology, 1-4, p. 57a.

Triebold, M., and Russell, D. A., 1996, The Sandy Site: Small Dinosaurs from the Hell Creek Formation of South Dakota: In: Dinofest International Symposium, programs and abstracts, edited by Wolberg, D. L., and Stump, E., Arizona State University, p. 109.

Wall, W. P., and Galton, P. M., 1979, Notes on pachycephalosaurid dinosaurs (Reptilia: Ornithischia) from North America, with comments on their status as ornithopods: Canadian Journal of Earth Sciences, v. 16, p. 1176-1186.

PREHISTORIC TIMES

Feb/March 1999   No. 34

THE PT
INTERVIEW:
JOHN
SIBBICK

ALSO
CADILLACS
&
DINOSAURS
MARK SCHULTZ

PALEONEWS
'98
ALL THE
TOP SCIENCE
STORIES OF
LAST YEAR!

PLUS:
BOOK &
MODEL REVIEWS
AND
MUCH MORE!

U.S. $5.95 • Canada $6.95

Ford, T. L., 1998, How to Draw Dinosaurs. *Albertosaurus/Gorgosaurus*, two peas in a pod? Or Apples and Oranges?: Prehistoric Times, n. 32, p. 12-13.

Chapter 12

Albertosaurus/Gorgosaurus, two peas in a pod? Or Apples and Oranges?

Besides *Tyrannosaurus rex*, *Albertosaurus* is the second famous tyrannosaurid known from North America. There is confusion between the names *Albertosaurus* and another tyrannosaur, *Gorgosaurus*. *Albertosaurus* was first described 1905 by Henry Fairfield Osborn in the same paper that *Tyrannosaurus rex* was described and was given the species name *sarcophagus*. The type is a fragmentary skull (National Museum of Nature, Ottawa, Ontario, Canada, NMC 5600; **Editors note: now known as the Canadian Museum of Nature**) (Figure 1) and the paratype (NMC 5601) is of a skull and fragmentary skeleton. *Albertosaurus arctunguis*, a synonym of *sarcophagus*, was described by Arthur Parks, in 1928, based on a fragmentary skeleton (Figure 2). Originally the type and paratype were described as Skull #2 and Skull #1 respectively and as *Laelaps incrassatus* (Cope, 1876). Unfortunately Parks died before the skull material could be illustrated. Lawrence Morris Lambe did described the referred skull material in 1902 to *Dryptosaurus incrassatus*. This is not to be confused with the type material of *Dryptosaurus incrassatus* teeth, which is from an early formation. Before this, *Laelaps* had been found to be preoccupied and renamed *Dryptosaurus*. Ken Carpenter (1992) drew for the first time the uncrushed skull of *Albertosaurus sarcophagus*. (Figure 3). The type specimen is from the Horseshoe Canyon Formation, (formerly Member B of the lower Edmonton Formation), Maastrichtian, Late Cretaceous and was found on the shore of the Red Deer River near Drumheller, Alberta, Canada. Several other specimens have been described as *A. sarcophagus*, many of which are at the Royal Tyrrell Museum of Paleontology in Drumheller, Alberta, Canada, and show how small the forelimbs are and short the skull is.

Gorgosaurus libratus was described by Lawrance Lambe, in 1914, 1917 (NMC 2120). It is based on a nearly complete skull and skeleton (Figure 4). One skull has set the standard for *Gorgosaurus* (under the name *Albertosaurus* by Dale Russell, 1970, Chicago Field Museum, FMNH PR308). *Gorgosaurus* was found in Dinosaur Provincial Park, Alberta, Canada, and is from the Dinosaur Park Formation, Judith River Group, Campanian, Late Cretaceous. Several other specimens ranging from Canada to New Mexico, have been found and attributed to *Gorgosaurus libratus*. There may be another species referred to *Gorgosaurus*, and that is *Gorgosaurus sternbergi*, which is a smaller thinner animal (Figure 5).

So are they the same? I can't comment on a lot of it because it is currently being studied by various (?) paleontologist. But here is what I can say. The reason *Albertosaurus* and *Gorgosaurus* were thought to be the same goes back to Dale Russell's 1970 paper on tyrannosaurids, and had been followed for years. New material of both genera has persuaded many paleontologist to now believe that they are actually different genera (which Dale Russell now agrees).

Basically, *Albertosaurus* is a larger, bulkier animal than *Gorgosaurus*. It has a shorter skull to height. There is a skull that Greg Paul (1988) referred to *A. sarcophagus* has a longer skull, which may be a male or different genus (Figure 6), with small lachrymal horns in the shape of small triangles with the apex of the 'horn' over the center of the lachrymal. The horn doesn't point forward. The teeth are smaller than *Gorgosaurus*, as is the neck and body. What is really interesting is that the forelimbs are shorter than *Gorgosaurus* and *Tyrannosaurus* relative to its body size. This means that *Tyrannosaurus* DIDN'T have the smallest forelimbs of the North American Tyrannosaurids! *Albertosaurus* is known just from the Maastrichtian age of the Late Cretaceous and mainly from the Horseshoe Canyon Formation.

Gorgosaurus libratus has been the stable tyrannosaur used by the name *Albertosaurus*. The main problem with this is the use of FMNH PR308 that Dale Russell used in his 1970 paper (Figure 7). As Tom Holtz has pointed out, PR308 is missing the top part of the skull! The lachrymal horn is triangular with some having the horn pointing forward. *Gorgosaurus* has a longer skull than *Albertosaurus* as well as neck, body and tail.

Gorgosaurus is restricted to the Campanian age of the Late Cretaceous, and has a range that extended deep into the United States, while *Albertosaurus* is known just from the Maastrichtian of Canada and was contemperous with *Tyrannosaurus rex*. Thus all the drawings, paintings, and sculptures that show a long skulled, thin bodied *Albertosaurus* from the Campanian are actually portraying *Gorgosaurus*.

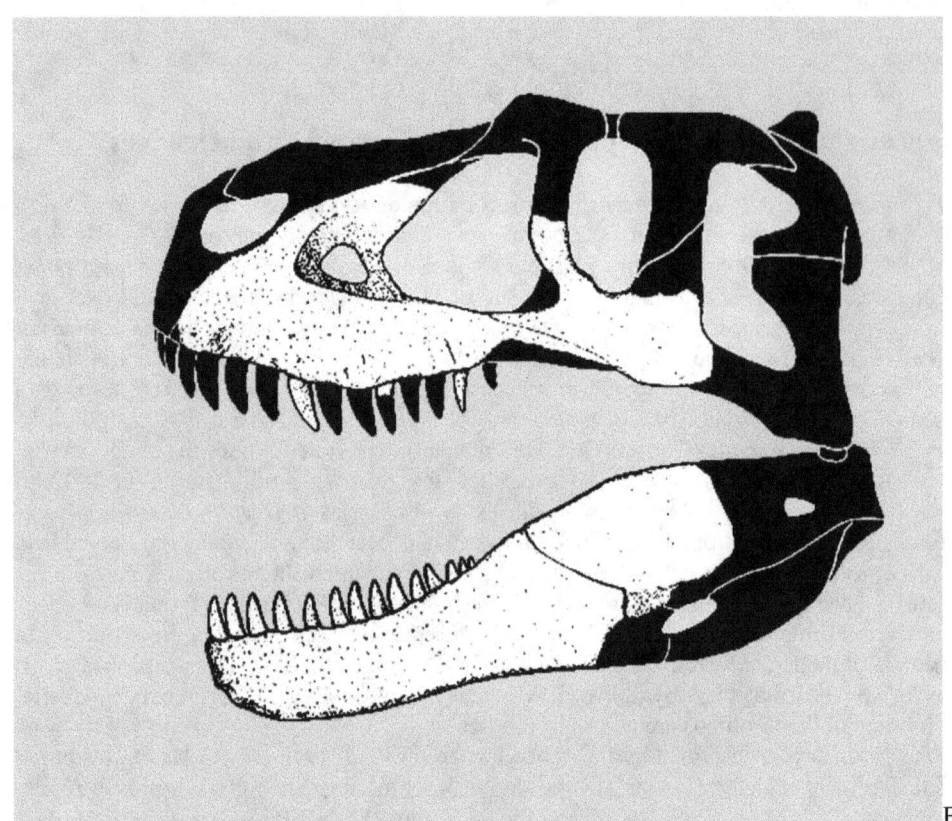

Figure 1), Skull of *Albertosaurus sarcophagus* type reconstructed to fit Ken Carpenters drawing.

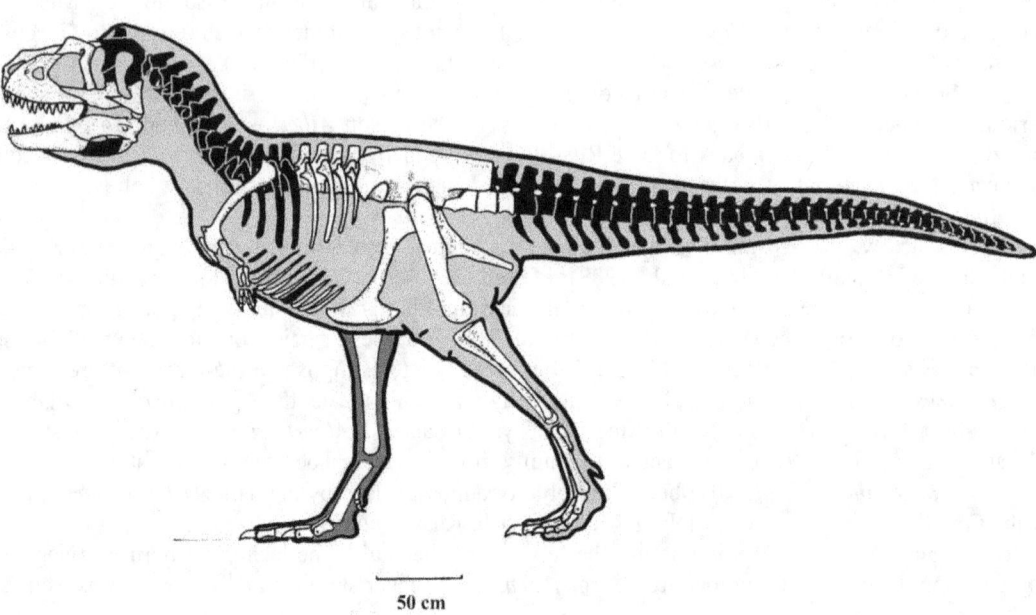

50 cm

Figure 2), Skeleton of *Albertosaurus arctunguis*. Note the small front legs.

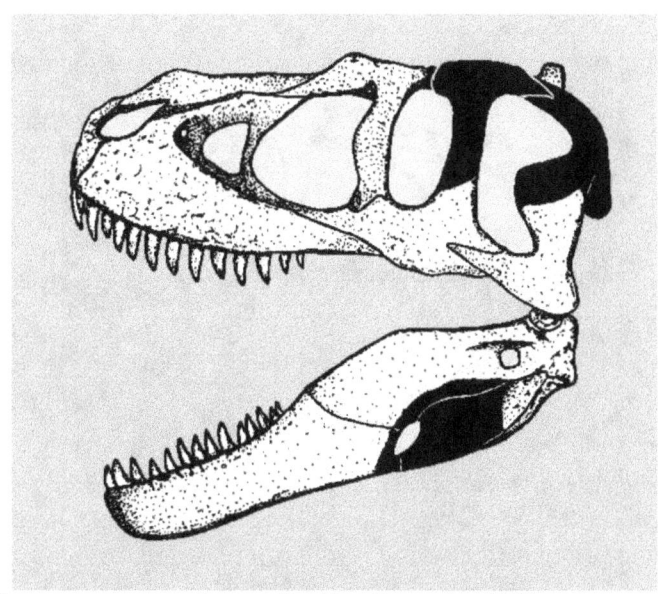

Figure 3), Skull of *Albertosaurus sarcophagus* after Ken Carpenter, 1992.

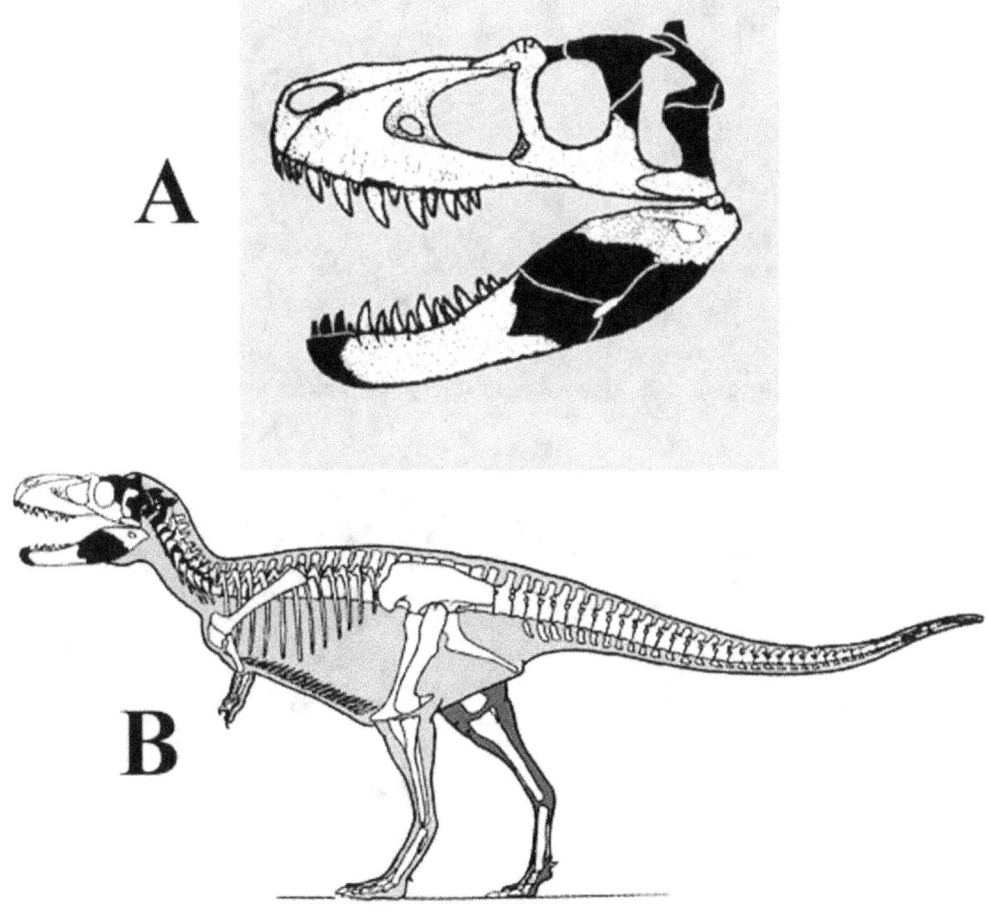

Figure 4), A) Skull, and B) skeleton of *Gorgosaurus libratus*

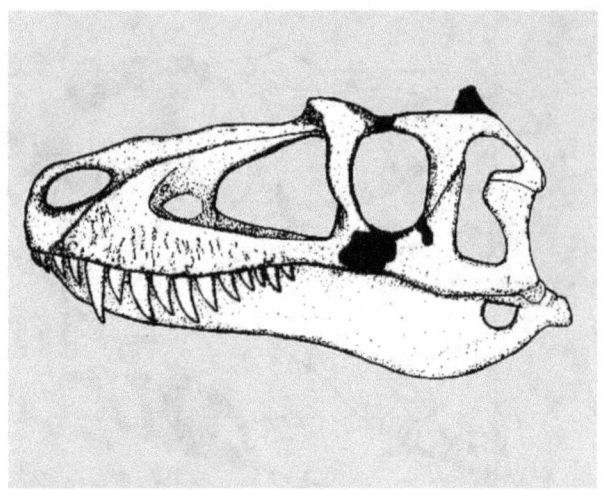

Figure 5), Skull of *Gorgosaurus sternbergi*.

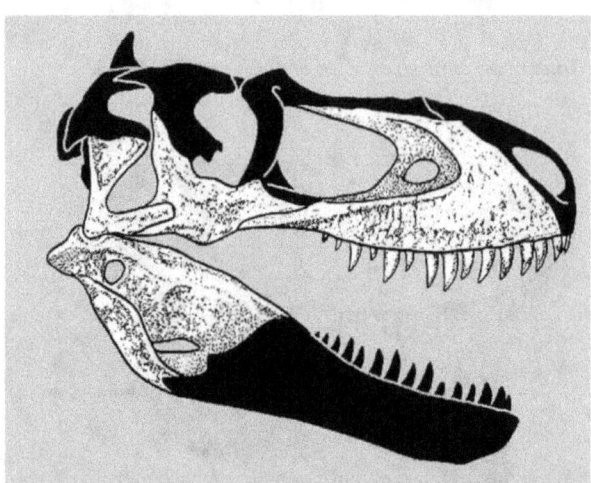

Figure 6), Skull of a referred long snouted *Albertosaurus sarcophagus* after Paul, 1988.

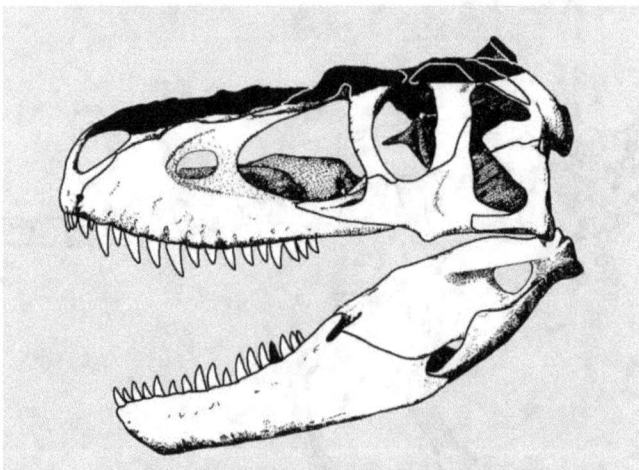

Figure 7), Skull of FMNH PR308, *Gorgosaurus libratus* showing the missing top of the skull.

Figure 8), Life restoration of; A) *Albertosaurus,* and B) *Gorgosaurus.*

Bibliography

Carpenter, K. 1992. Tyrannosaurids (Dinosauria) of Asia and America. In: Aspects of Nonmarine Cretaceous Geology. China Ocean Press: 250-268.

Cope, E. D. 1892. On the Skull of the dinosaurian *Laelaps incrassatus* COPE. Amer. Phil. Soc. Proc. Vol. 30: 240-245.

Lambe, L. M. 1902. 2. New genera and species from the Belly River Series (Mid-Cretaceous). Contributions to Canadian Palaeontology, Vol. III. Geol. Surv. Can: 22-81.

Lambe, L. 1904. On *Dryptosaurus incerassatus* (COPE) from the Edmonton Series of the North West Territory. Geol. Surv. Canada, Contrib, Canadian Paleont., Vol. 3, Part 3: 1-27.

Lambe, L. 1914. On a new genus and species of carnivorous Dinosaur from the Belly River Formation of Alberta, with description of the skull of *Stephanosaurus marginatus* from the same Horizon. Ottawa Naturalist, Vol. 28: 13-20.

Lambe, L. 1917. The Cretaceous theropodous dinosaur *Gorgosaurus*. Geol. Surv. Canada Memoir, Vol. 100: 1-84.

Olshevsky, George, 1996. *Albertosaurus*. The Dinosaur Folios, First Installment: May 1996: 14pp.

Osborn, H. F. 1905. *Tyrannosaurus* and other Cretaceous carnivores Dinosaurs. Bull. Amer. Mus. Nat. Hist., 21: 259-265.

Parks, W. A., 1928. *Albertosaurus arctunguis*, a new species of theropodous dinosaur from the Edmonton Formation of Alberta. University of Toronto Studies, Geology Series, Volume 25: 1-42.

Paul, G. S. 1988. Predatory Dinosaurs of the World, a complete Illustrated guide. New York Academy of sciences book: 464pp.

Russell D. A. 1970, Tyrannosaurs from the Late Cretaceous of Western Canada. Natl. Mus. Nat. Sci. Pub. Pal. No. 1. 1-34.

PREHISTORIC TIMES

DEC / JAN '98/99   NO.33

PT INTERVIEWS:
DOUGLAS HENDERSON
KAREN CHIN

DINOSAUR MODEL
AND BOOK REVIEWS
PLUS MUCH MORE!

U.S. $5.95 • Canada $6.95

Ford, T. L., 1998-1999, How to Draw Dinosaurs. How low does the body go?: Prehistoric Times, n. 33, p. 12-13.

Chapter 13

How low does the body go?

One of the things that I look for when I look at a drawing, painting or sculpture of a theropod is how deep the artist made the body and the pelvis and how to he portrayed the body at the knee. This is something that many artist get wrong. They make the body too shallow at the knee. This is accomplished in two ways, how deep the pubis is and how long the femur is in relation to the pelvis.

The pelvis is make up of the ilium (top), pubis (front) and ischium (back) (figure 1). The ilium varies in height, length, and in how wide the top of the ilia is (plural of ilium, plural of pubis is pubi, and ischium is ischia). In *Tyrannosaurus* the ilia touch each other, where in other theropods there is a gap.

In the pubis, the lower end is either a small rod or expands into a boot. For example in allosaurids, and tyrannosaurids the pubic boot is very large. The pubis varies greatly in length, angle and shape. In ceratosaurs the pubis is angled forward and in *Coelophysis* (Figure 2a) the end is rod-like. With *Ceratosaurus*, (figure 2b) it is slightly expanded. The knee is blow the pubis in both genera. *Compsognathus* (figure 2c) has a more vertical pubis with a slightly expanded boot pointing backwards and the knee is slightly below the pubis. Oviraptorid *Ingenia* (figure 2d) has a strange pubis, it curves forward on the lower end and has a slight expansion in front, while the knee is below the pubis. The ornithomimid *Dromiceiomimus* (figure 2e) has near vertical pubis with an expanded hatchet end of the boot with the knee at the end of the pubis. *Allosaurus* (figure 2f) has a near vertical pubis, with an extremely expanded boot facing backwards and slightly upwards and the knee is at the same level of the pubis. *Tyrannosaurus* (figure 2g) has a pubis that slightly curves forward and has one of the largest pubic boots of any theropod. The knee is at the same length as the pubis. The pelvis of dromaeosaurs are very bird like. In *Velociraptor* (figure 2h) the pubis faces backwards and has a large pubic boot. The pubis extends beyond the back of the ilium and the knee is lower than the pubis (in fact the drawing shows the pubis at too shallow of an angle. Mark Norell gave a talk at the SVP) that showed the pubis at much higher angel. When the paper comes out, it will change how dinosaurs and birds are looked at).

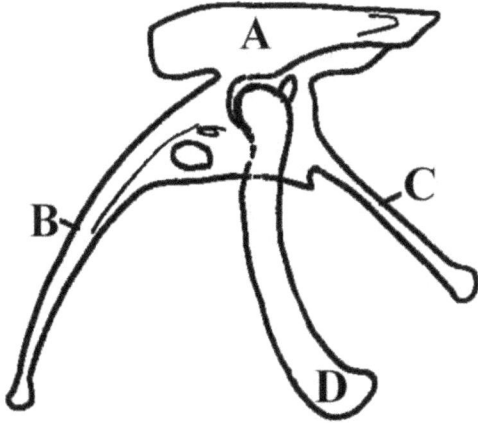

Figure 1), Pelvis of *Coelophysis* (after Row and Gauthier, 1990); A) ilium; B), pubis; C) ischium, and D) femur.

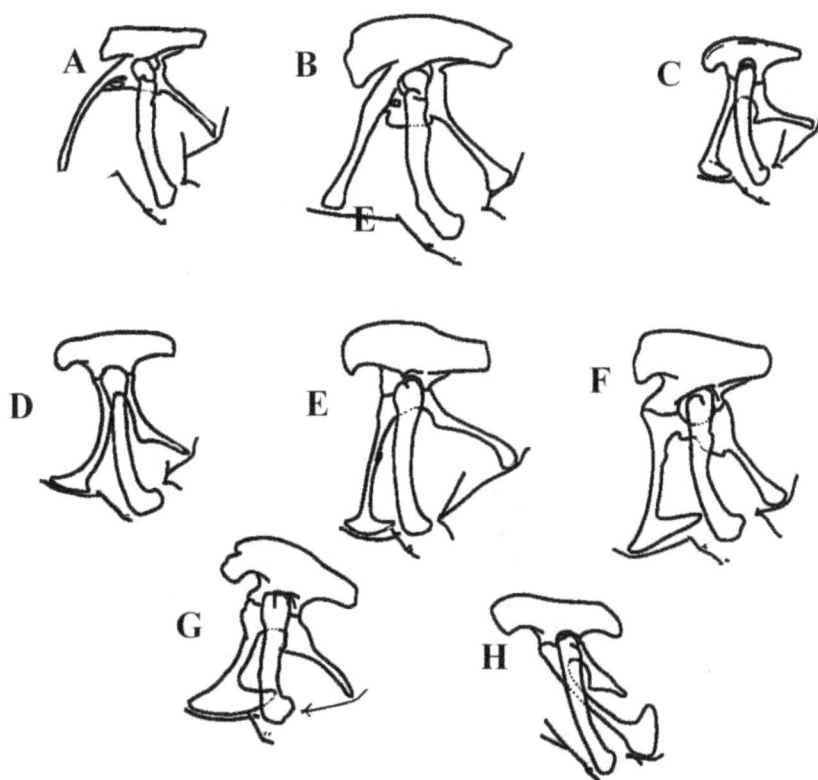

Figure 2), Pelvis and femur of theropods showing the position of the knee and body; A) *Coelophysis* (after Paul, 1996); B) *Ceratosaurus* (after Paul, 1996); C) *Compsognathus* (after Norman, 1990); D) *Ingenia* (after Barsbold, Maryanska and Osmolska, 1990); E) *Dromiceiomimus* (afer Barsbold & Osmolska, 1990); F) *Allosaurus* (after Paul, 1996); G) *Tyrannosaurus* (after Paul, 1996); and H) *Velociraptor* (after Paul, 1996).

Bibliography.

Barsbold, R., Maryanska, T., and Osmolska, H., 1990, Oviraptorosauria: In: Dinosauria, edited by Weishampel, D. B., Dodson, P., and Osmolska, H., California University Press, p. 249-258.

Barsbold, R., and Osmolska, H., 1990, Ornithomimosauria: In: Dinosauria, edited by Weishampel, D. B., Dodson, P., and Osmolska, H., California University Press, p. 225-244.

Norman, D. B., 1990, Problematic Theropoda: "Coelurosaurs": In: Dinosauria, edited by Weishampel, D. B., Dodson, P., and Osmolska, H., California University Press, p. 280-306.

Paul, G., 1996, The complete illustrated guide to Dinosaur skeleton: Gakken Mook, 97pp.

Rowe, T., and Gauthier, J., 1990, Dinosaur Taxonomy, Part II. Ceratosauria: In: Dinosauria, edited by Weishampel, D. B., Dodson, P., and Osmolska, H., California University Press, p. 151-168.

For the Dinosaur Collector and Enthusiast

PREHISTORIC TIMES

Oct - Nov 1998   No.32

Messmore & Damon's

Mechanical Dinosaurs

The PT Interview
Artist: Mark Hallett

PREHISTORIC SHARKS

Reviews of all the latest dinosaur items

And Much Much More

U.S. $5.95 • Canada $6.95

10>

0 74470 92325 1

95

Ford, T. L., 1999, How to Draw Dinosaurs. How many fingers did *Compsognathus* have?: Prehistoric Times, n. 34, p. 14-16.

Chapter 14

How many fingers did *Compsognathus* have?

Compsognathus has an unique problem, one that some believe to be solved, yet one that I believe needs to be addressed once again. That is, how many fingers did it have? An answer to the question will aid in theropod reconstructions.

This famous fossil has intrigued many a paleontologist and layman alike for the last 137 years. Andreas Wagner first described the specimen in 1861 with very little fanfare. It wasn't until Thomas Huxley description of it being a "bird-like reptile" that it started to gain its fame. This small theropod, to Huxley's surmise, showed many bird-like attributes; long legs, hollow bones, a slender bird-like skull etc. Many have looked over the specimen and a 'normal' three-finger hand to it. In 1978 John Ostrom published a monograph on *Compsognathus,* and concluded, after looking over the specimen with a fine tooth comb, so to speak, that it had only two fingers, quite unlike 'normal' theropods for that time period. A two digit hand had only been known in tyrannosaurids. Is John Ostrom right? Or is the specimens hands so crushed and disassembled to be unable to tell? Other specimens are needed to help to come to a correct conclusion. These will be addressed shortly. First we need to understand what different theropod hands look like.

A typical theropod hand consist of metacarpals, phalanges and unguals. Metatarsals are the bones that make up the fleshy part of the hand and do not move. The phalanges are the fingers; i.e. the thumb (digit 1), and one to 4 fingers (digits 2-5). Digit one (I won't be taking into account that digit one may be replaced by digit two as Gauthier has proposed to coincide with birds, and try as to not confuse the reader) has only one phalange, and claw (ungual), digit two has two phalanges, digit three has three, digit four has four and digit five has five (figure 1). This is true only when all the digits have claws. This arrangement is true in all the theropods that I know of. It differs in other types of dinosaurs. Some have two digits, others have five.

The earliest theropod (I will try and refrain from using the terms primitive and advanced due to the confusion to some this may cause) *Herrerasaurus* (figure 2a) has five digits with digit three being the longest, quite unlike later theropods. Digit four and five are nearly non existent. *Syntarsus* has three fingers (figure 2b), with a nearly non existent forth. Even though *Ceratosaurus* is known only from a fragmentary hand and drawn incorrectly with 4 expanded metacarpals (corrected in figure 2c).

Carnotaurus also has a fragmentary hand. It is very unusual (figure 2d). It has four metacarpals with the fourth being a spike like element. *Allosaurus* has three digits (figure 2e). *Struthiomimus* has three digits with the first one being very long (figure 2f). *Oviraptor* has three digits (figure 2g). *Ingenia* has much shorter digits (figure 2h). *Deninonychus* has three digits (figure 2i). *Albertosaurus* has 2 digits (figure 2j). Let me clarify this last.

Tyrannosaurids do have three metacarpals, with metacarpal three just lying next to metacarpal two. This actually will in no way change any reconstruction of tyrannosaurids, but it is interesting to note. *Ornitholestes*, even though it has three digits, digit one missing beyond the metacarpal, not due to it having only two digits, due to poor preservation (Figure 2k). *Saurornithoides* has three digits with digit one able to become opposable (Figure 2l), like a human thumb.

Because digit one has only one phalange; it would have the least arch of movement when the fingers are grasping. It would also, in some cases, move offset from the other digits. Because digit three has the most phalanges, it would have the most movement. (figure 3). The claws could also move and aid in capture, etc. If you prefer that the fingers be straight and just move the claws, this could also happen. Not all fingers would close to the same degree. Some could clench more than others. I think this needs to be looked at by a paleontologist or someone who wants to seriously study this; to determine how far different theropods could clench their fists, so to say.

How does *Compsognathus* fit in? If Compy were a typical theropod it should have three fingers, but the type specimen (*Compsognathus longipes*, figure 4a, 5a) has four phalanges and claws, two for each hand. (figure 6a). Also, unlike other theropods digit two, according to Ostrom, only has one phalange which differs greatly from all other theropods.

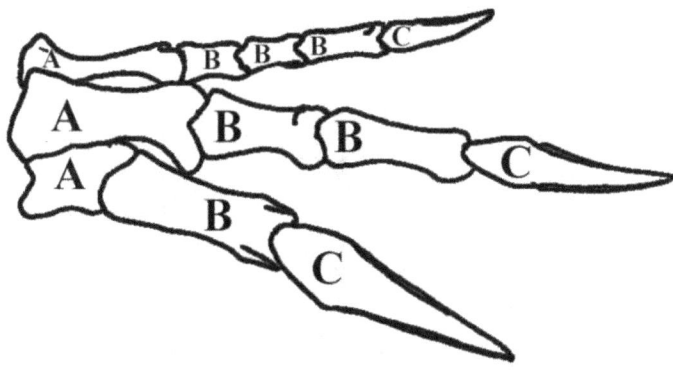

Figure 1), *Allosaurus* hand showing metacarpals (A); phalanges (B); and claws (C). (after Madsen, 19)

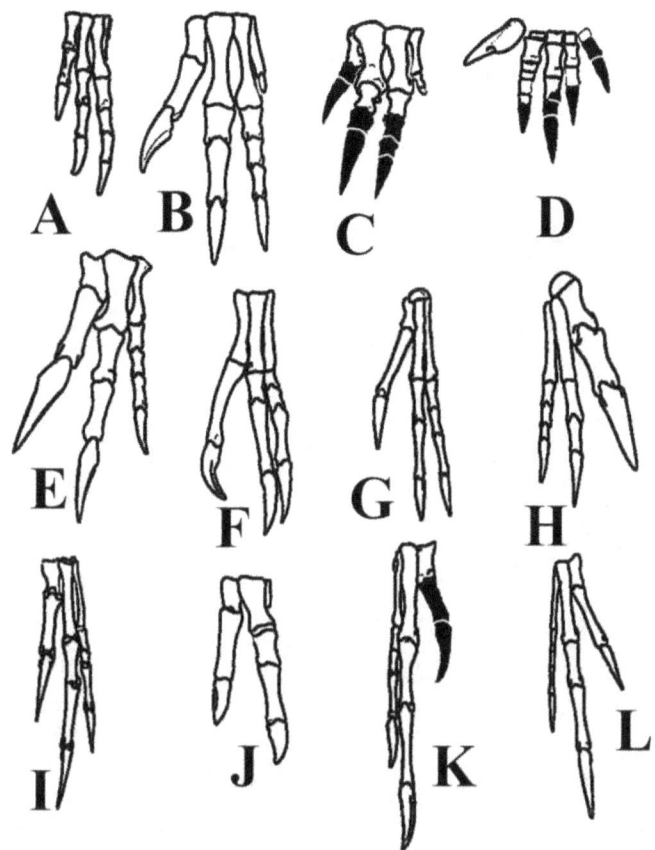

Figure 2), A) *Herrerasaurus* hand (after Sereno, 1993); B) *Syntarsus* (after Rowe & Gauthier, 1990); C) *Ceratosaurus* (modified after Gilmore, 1920); D) *Carnotaurus* (after Bonaparate et al, 1990); E) *Allosaurus* (after Madsen, 1976); F), *Struthiomimus* (after Barsbold & Osmolska, 1990); G) *Oviraptor* (after Barsbold et al, 1990); H) *Ingenia* (after Barsbold et al, 1990   ); I) *Deninonychus* (after Ostrom, 1990.); J) *Albertosaurus* (after Molnar et al,1990); K) *Ornitholestes* (after Molnar, 1990); L) *Saurornithoides* (after Russell, 1969).

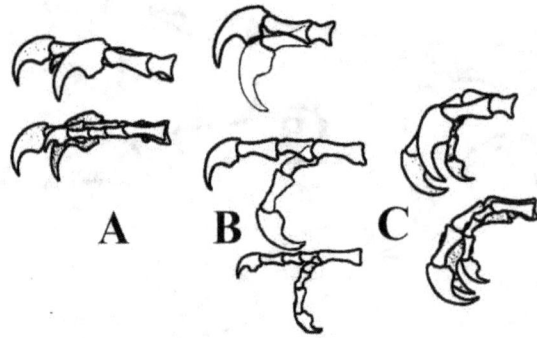

Figure 3), Hand of *Allosaurus* in side view with digit 1 in front view, with digit 3 in front A); showing the arch of the fingers and claws B); hand clutching with digit 1 in front and digit 3 in front. This is by no means the excat extent that the hand could close, it just gives the reader an idea on how the fingers close.

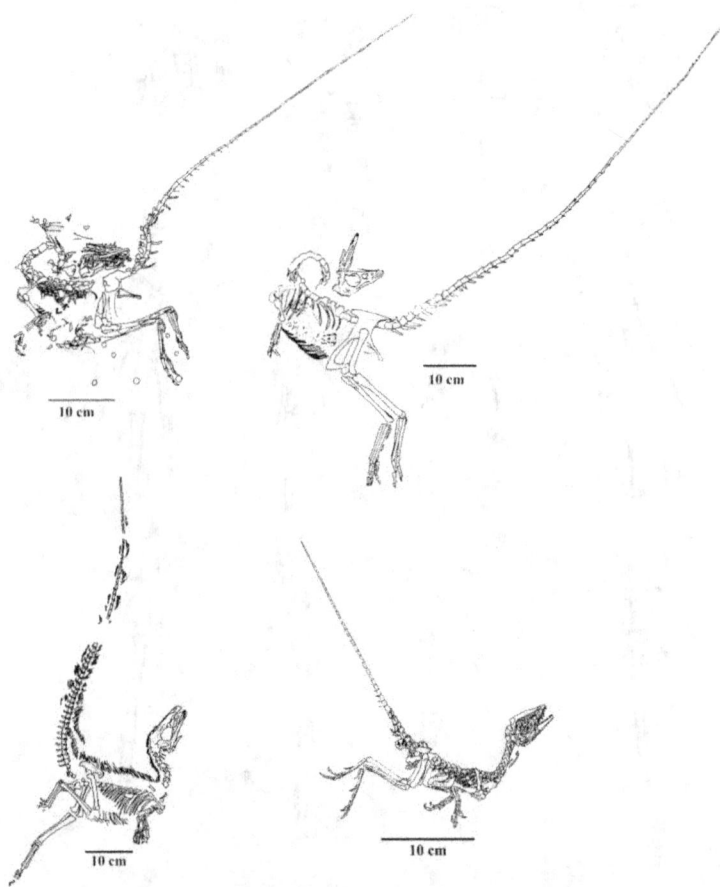

Figure 4), Small theropod skeletons in situ A). *Compsognathus longipes* (after Ostrom, 1978); B). *Compsognathus corallestris* (after Bider, Demay & Thomel, 1972); C). *Sinosauropteryx* (after Chen, Dong & Zhen, 1998); D). *Scipionyx* hand (after Dal Sasso and Signore, 1998).

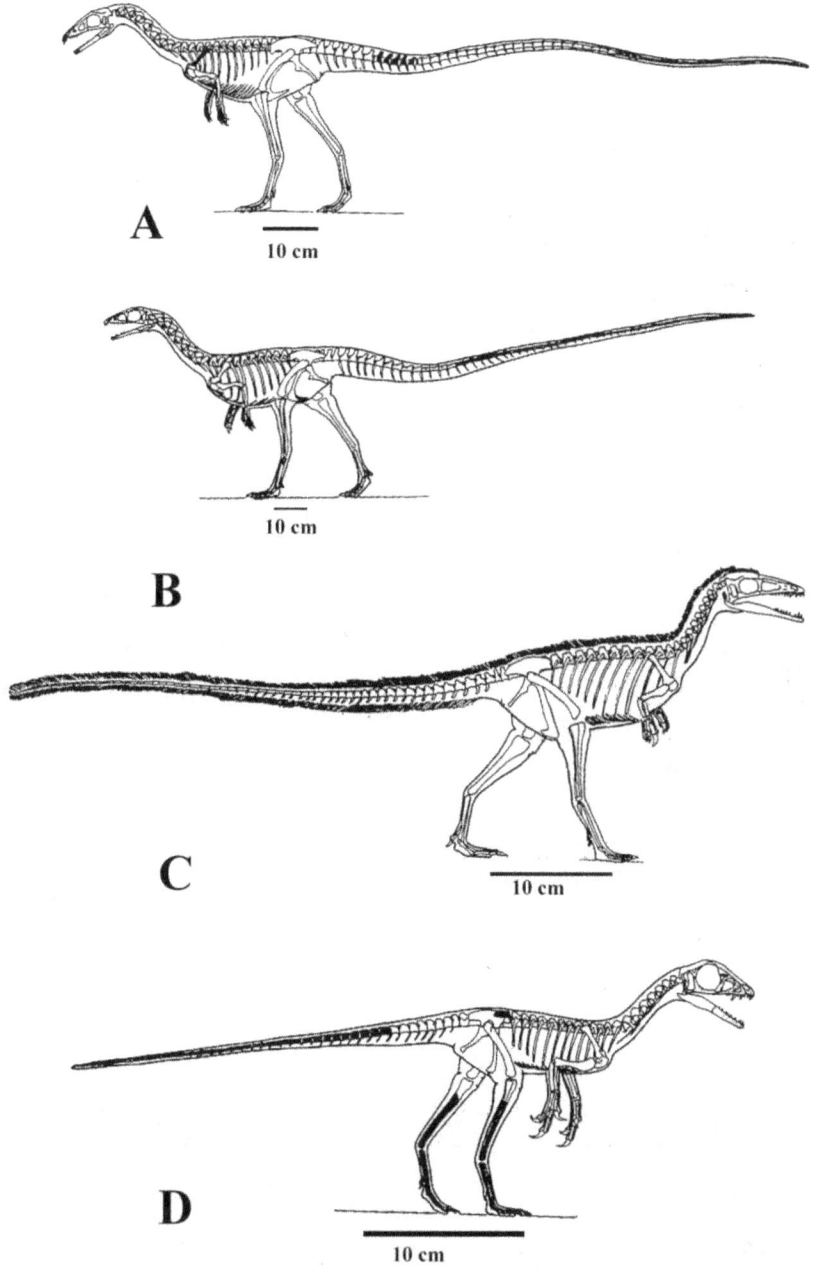

Figure 4), Small theropod skeletons in situ A). *Compsognathus longipes* (after Ostrom, 1978); B). *Compsognathus corallestris* (after Bider, Demay & Thomel, 1972); C). *Sinosauropteryx* (after Chen, Dong & Zhen, 1998); D). *Scipionyx* hand (after Dal Sasso and Signore, 1998).

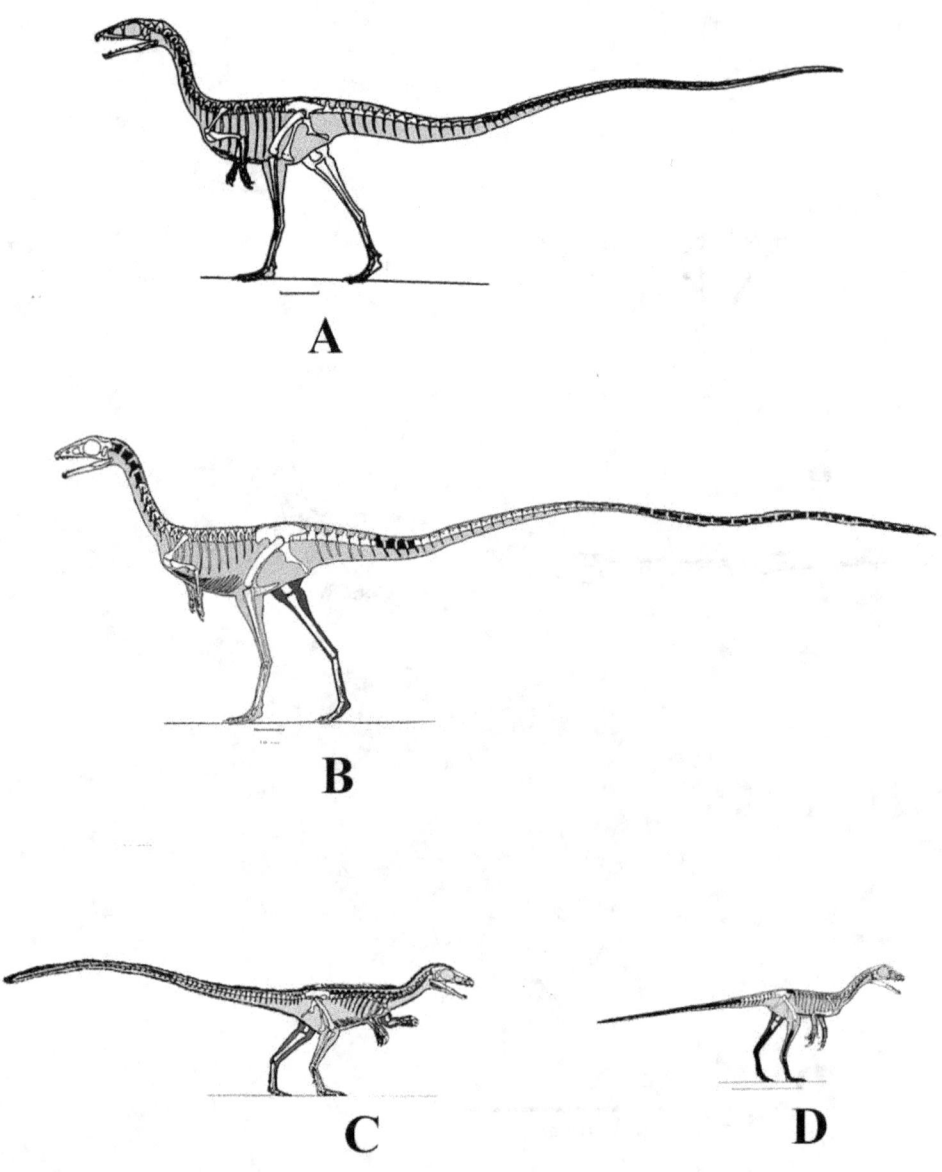

Figure 5). Small theropod skeletons reconstructed A). *Compsognathus longipes* (after Ostrom, 1978); B). *Compsognathus corallestris* (after Bider, Demay & Thomel, 1972); C). *Sinosauropteryx* (after Chen, Dong & Zhen, 1998); D). *Scipionyx* hand (after Dal Sasso and Signore, 1998).

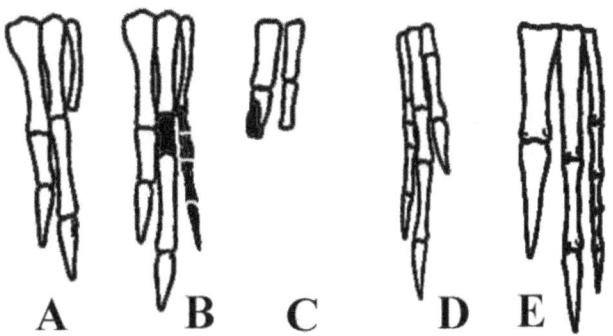

Figure 6), A). A two fingered hand of *Compsognathus longipes* (after Ostrom, 1978), and showing how it would look if it had 3 fingers. B); *Compsognathus corallestris* (after Bider, Demay & Thomel, 1972); C). *Sinosauropteryx* (after Chen, Dong & Zhen, 1998). D). *Scipionyx* hand (after Dal Sasso and Signore, 1998).

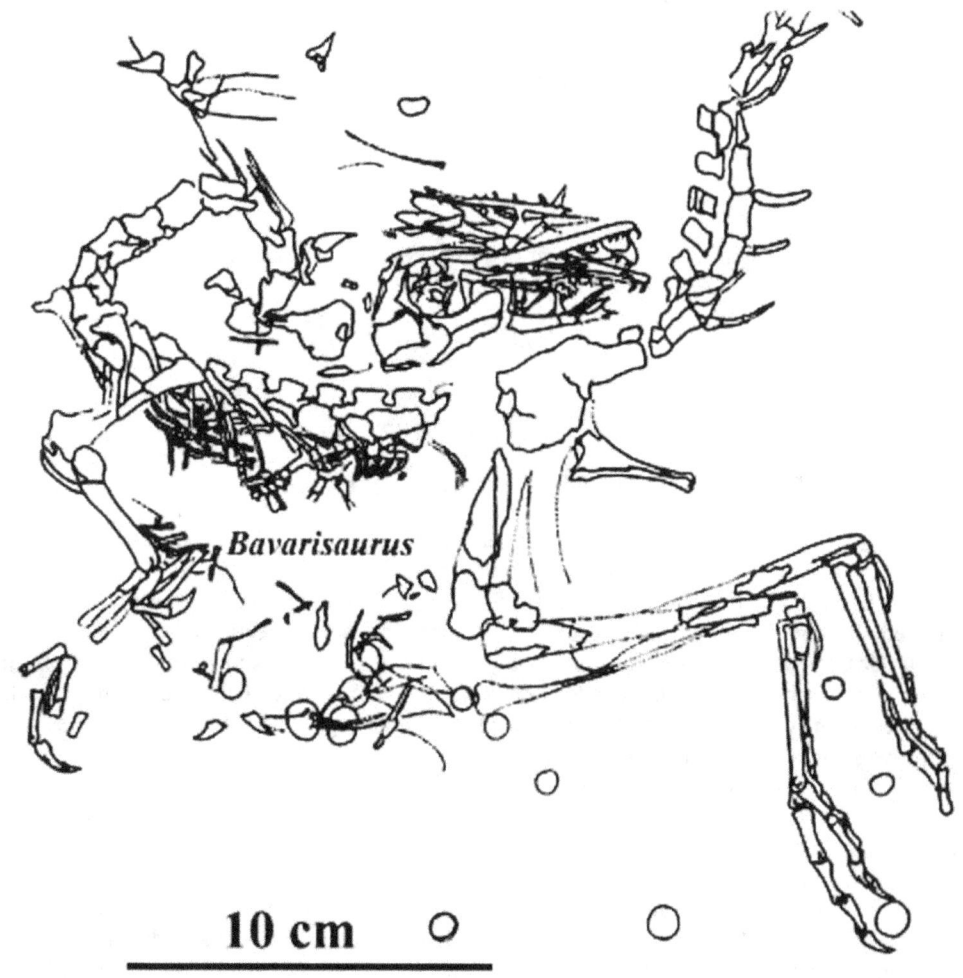

Figure 7), *Compsognathus longipes* skeleton showing the folded up skeleton of *Bavarisaurus* in the body cavity (L), and the 13 eggs (round objects).

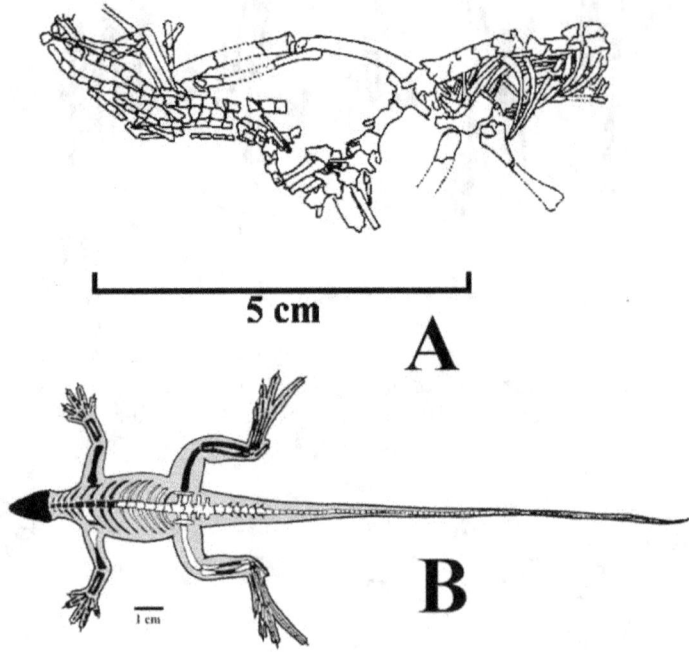

Figure 8), *Bavarisaurus* folded skeleton (after Ostrom, 1978) (A), and skeleton reconstruction (B).

Figure 9), *Saccocoma* general drawing (after Frickhinger, 1994)

What about the French specimen? The hands of *C. corallestris* (Bider, Demay & Thomel, 1972, Figure 4b, 5b) are unfortunately nearly totally missing (figure 6b). At the time Bider et al thought *C. corallestris* had flippers for hands and that it was an aquatic animal. Ostrom interpreted this 'flipper' to be inaccurate and improbable, which has been the interpretation, and correctly so, ever since. *Sinosauropteryx* (figure 4c, 5c), even though similar and possibly a compsognathid has most of its hand preserved (in the type specimen, figure 6c). Then there is *Scipionyx* (figure 4d, 5d) which has the best preserved hands of any of the talked about small theropods. It clearly has three digits (figure 6d). At one time it was speculated that *Scipionyx* was a compsognathid, but no longer.

So, how many fingers does *Compsognathus* have? I would be inclined to say it had three, and that the hands were just not preserved well enough to prove it. Two fingers may be incorrect for *Compsognathus*, but that is just one of the things that intrigues me about paleontology, not everything is cut and dry, which leads to many an interesting conversation.

Now I'd like to bring up a few more things about the type specimen of *Compsognathus* that not everyone knows. For decades it was thought that *Compsognathus* was either a cannibal or gave live birth due to a small skeleton in the stomach region. Ostrom showed that it was its last meal. A small lizard *Bavarisaurus macrodactylus*. *Bavarisaurus* was most likely a swift long tailed lizard (figure 7, 8.)

Peter Griffiths published a paper in 1993 on the type specimen having 13 eggs. Small, spherical and flattened objects are found in and around the body cavity. The presence of eggs in a body cavity doesn't mean that the animal was an adult, only that it was reproductively active female. For instance, teen aged girls can have babies. *C. longipes* could be a "teenager" and *C. corallestris* can be an adult. What the eggs do indicate is that *C. longipes* is a female, and Griffiths believes *C. corallestris* to be a male. If this is so, and if *C. corallestris* is a synonym of *C. longipes*, than *C. longipes* is based on a 'teenage' animal. I base this conclusion on birds where the female is larger than the males. If *C. longipes* is an adult female, than it wouldn't be the same species as *C. corallestris* because it is much smaller than *C. corallestris.*

Ostrom also was interested in looking at various theories regarding *Compsognathus,* like whether or not von Huene was correct in *C. longipes* having 'skin-armor' or Nopcsa's observations on "parallel and irregular fibers". Both Nopcsa and Ostrom dismiss Huene's "skin-armor" as being an artifact of preservation. Ostrom does believe that there may be "parallel and irregular fibers" along the lower edge of the ischium, between the right radius and ulna, and very faint lineation along the dorsal region of the proximal caudals close to the tip of the right dentary. This maybe indications of soft tissue, and should be looked at further.

Both specimens of *Compsognathus* has most of the tail missing, and are presumed to be of average length. *Sinosauropteryx* specimens have a very long tail. If *Compsognathus* and *Sinosauropteryx* are similar to each other, it is possible that the tail of *Compsognathus* was just as long. By using the length of the caudal vertebrae of *Compsognathus corallestris* and comparing the same caudal vertebra length of *Sinosauropteryx* a new length of the tail of *Compsognathus* can be made. This will only work if the corresponding caudal vertebrae have the same percentage length to one another. I've done what I can and they do seem to be proportionally the same. I've also modified the skeleton of *Compsognathus* to accommodate this. (This is just my opinion and may not be the correct one.)

One last thing about *C. longipes*. It is not totally clear where the specimen is from. It could either be from Kelheim, as suggested by a hand written label on the underside of the slab or Jachenhausen, as suggested by a typed label, the author of which is not known. Ostrom discounted Jachenhausen due to *Saccocoma* being found on the slab, which is not known from Jachenhausen. *Saccocoma* is a free swimming sea lily (figure 9). Griffiths does not believe that there are remains of *Saccocoma*, that the specimen is from Jachenhausen.

Bibliography

Barsbold, Rinchen, Teresa Maryanska and Halzska Osmolska, 1990. Oviraptorosauria. In: Dinosauria, Edited by David B. Weishampel, Peter Dodson and Halska Osmolska. California University Press: 249-258.

Barsbold , Rinchen, and Halszka Osmolska, 1990. Ornithomimosauria. In: Dinosauria, Edited by David B. Weishampel, Peter Dodson and Halska Osmolska. California Univeristy Press: 225-244.

Bidar, A., Demay, L., and Thomel, G. 1972. Sur la presence du Dinosaurien *Compsognathus* dans le Portlandien de Canjuers (Var). C. R. Acad. Sc. Paris, t. 275, Serie D: 2327-2329.

Bidar, Alain, Louis Demay and Gerard Thomel, 1972. *Compsognathus corallestris* Nouvelle Espece de Dinosaurien Theropode du Portlandien de Canjuers (Sud-est de la France). Extrait des Annales du Msueum d'Histoire Naturelle de Nice, Tome 1, Fascicule 1: 3-34.

Bonaparte, J. F., Novas, F. E., and Coria, R. A. 1990. *Carnotaurus sastrei* BONAPARTE, the Horned, Lightly Built Carnosaur from the Middle Cretaceous of Patagonia. Contributions in Science, Number 416: 1-41.

Chen, Pei-je, Zhim-ming Dong and Shuo-nan Zhen, 1998. An exceptionally well-preserved theropod dinosaur from the Yixian Formation of China. Nature, Volume 391: 147-152.

Dal Sasso, Cristiano, and Marco Signore, 1998. Exceptional soft-tissue preservation in a theropod dinosaur from Italy. Nature, Volume 392: 383-387.

Frickhinger, K. A., 1994. Die Fossilien von . The Fossils of Solnhofen. Goldschneck-Verlag: 336pp.

Gilmore, C. W. 1920. Osteology of the Carnivorous Dinosauria in the United States National Museum, with special reference to the genera *Antrodemus* (*Allosaurus*) and *Ceratosaurus*. Bull. U. S. Nat. Mus. No. 110: 1-159.

Griffiths, Peter, 1993. The question of *Compsognathus* eggs. Revue de Paleobiologie, Volume special number 7: 85-94.

Ji, Qiang et al., 1996. On discovery of the earliest bird fossil in China and the origin of birds. Chinese Geology, 1996. 10, No. 233: 30-33.

Madsen, J. H. jr. 1976. *Allosaurus fragilis* a revised osteology. Utah Geological and Mineral Survey, Bull. 109: 1-163. Reprinted 1993.

Molnar, Ralph E., 1990. Problematic Theropoda: "Carnosaurs". In: Dinosauria, Edited by David B. Weishampel, Peter Dodson and Halska Osmolska. California University Press: 307-317.

Molnar, Ralph E., Seriozha M. Kurzanov and Dong Zhiming, 1990. Carnosauria. In: Dinosauria, Edited by David B. Weishampel, Peter Dodson and Halska Osmolska. California University Press: 169-209.

Ostrom, J. H. 1978. The Osteology of *Compsognathus longipes* WAGNER. Zittelliana, Vol. 4: 73-118.

Ostrom, John H., 1990. Dromaeosauridae. In: Dinosauria, Edited by David B. Weishampel, Peter Dodson and Halska Osmolska. California University Press: 269-279.

Rowe, Timothy and Jacques, Gauthier, 1992. Dinosaur Taxonomy, Part II. Ceratosauria. In: Dinosauria, Edited by David B. Weishampel, Peter Dodson and Halska Osmolska. California University Press: 150-168.

Russell, D. A. 1969. A new specimen of *Stenonychosaurus* from the Oldman Formation (Cretaceous) of Alberta. Can. Jour. Earth Sci. Vol. 6: 595-612.

Sereno, Paul C. 1993. The pectoral girdle and forelimb of the basal Theropod *Herrerasaurus ischigualastensis*. Journal of Vertebrate Paleontology, Volume 13, Number 4: 425-450.

Wagner, J. A. 1961. Neue Beitrage zur Kenntnis der urweltlichen Fauna des Lithographischen Schiefers. II. Schildkorten und Saurier. Abh. Bayer. Akad. Wiss, vol. lIX: 65-124.

For the Dinosaur Collector and Enthusiast

PreHistoric Times

Aug-Sep 1998   No. 31

The PT Interv   Paleontologist Paul Sereno

Tracy Ford's
How To Draw
Dinosaurs:
Pachycephalosaurus

Expert Dino
Modeling Tips

Sculpting
Prehistoric
Mammals

lus
Airfix
British Dinosaur
Model Kits
All the latest
Paleo News
Reviews of the lates
Dino kits
and much more!

U.S. $5.95 • Canada $6.95
08>

0 74470 92325 1

Ford, T. L., 1999, How to Draw Dinosaurs. Walk which way? Prehistoric Times, n. 35, p. 14-15.

Chapter 15

Walk which way?

The two famous hadrosaur mounted skeletons at the American Museum of Natural History of *Edmontosaurus*, (at first *Trachodon*, then *Anatosaurus*) defined how ornithopods were thought to have walked. One stood on its hind legs towering over the other walking on all fours looking for something to eat. This interpretation changed during the big dinosaur renaissance in the late 1960's. Scientist determined that the body should be posed horizontal with the focal point or balancing point at the location where the hind legs attached to the pelvis. The new interpretation portrayed the front legs off the ground and the animal walking bipedally. Then slowly a different slant on how ornithopods walked, came into view; they because quadrapedal (figure 1). Now with the use of footprints, paleontologist know that ornithopods were bipedal animals that walked occasionally on all four.

Science knows that the hind legs are much larger than the front and the hind feet have either 3 or 4 toes. In iguanodontids and hadrosaurids the front feet/hands have 4 toes with hoofed claws on the first two fingers, the last two lacking a claw entirely. The last finger was held away from the bundle of the first three fingers. Iguandontids have a spike like thumb claw.

As noted before, when ornithopods walked using the hind legs only, the man balance of weight was at the hip joint, causing the dinosaur to because a virtual, walking teeter-totter. The hind legs were also held directly under the animal with the hind feet almost aligned with the centerline of the body. When walking on all fours, the front feet were slightly outside the centerline. The hands faced forward and were in front of the hind feet. This is how iguanodontids were thought to have walked also.

Joanna Wright is challenging this theory. She has reexamined three trackways from Suttle's 'Mutton Hole Quarry, southwest of Herston in Dorset, England that are from the Intermarine Member of the Purbeck Limestone Group, Early Cretaceous. Here there are three trackways showing both manus (hand) and pes (hind foot) tracks. Trackway A has 14 pes tridactyle pes prints only, trackway Ca has 26 pes and 14 manus prints while Cb ahs 25 pes and 14 manus prints while Cb has 25 pes and 1 manus print. An interesting not is that trackways Ca and Cb were at first thought to belong to a single individual that was walking very slowly with a wide stance. After more of the trackways were uncovered, the tracks diverged and went their own ways.

These tracks show that the hind feet were directly under the body and the toes were widely splayed apart (figure 2a). Also, along the outside of Ca there are strange marks, that are identical, but reversed on both sides (figure 2b). Wright compared them to different *Iguanodon* hands. The digits are tightly pressed to one another., as in hadrosaurs, and have a hoof-like ungual (claw) on digit one and a larger one on digit two, and none on digit three, with digit 4 being free (figure 2c).

She concluded that the manus prints fit well with *Iguanodon*. Though there was no exact match to any of the *Iguanodons* or iguanodontids, the overall appearance indicates that they do belong to an iguanodontid.

The ulna and radius didn't cross one another and if the manus was held like typical hadrosaur the arms would have been disarticulated. Wright gave a talk about this at the Early-Middle Cretaceous Symposium late last year and I asked David Norman (without a doubt the one person who knows Iguandontids inside and out) if the skeletal elements supported laterally placed hands and he said yes. I've drawn Iguandontids with the thumb facing forward only to show off the thumb 'spike' (see Prehistoric times, issue 24, page 30: **Chapter four of this book**). It turns out that my drawings are correct in a round about way. It has been suggested that the manus was held in a large pad with the fingers unable to move. The trackways do not support this.

Hadrosaur hands are similar to iguanodon in that the fingers are held tightly together and have a free number 4 digit, but they lack the thumb spikes (figure 3a). Trackways show that the legs were held almost directly under the body with the manus prints lightly offset and in front of the pes. The pes shows that the toes were not spread as that of the iguandontid tracks (figure 3b).

As Wright has shown, the manus trackways differs from one another. Iguandontid manus prints have a strange shape with the manus perpendicular to the body (figure 4a). *Caririchnium* has rounded manus prints that may also have been perpendicular to the body (figure 4d). Hadrosaurs have small oval shaped manus that are in front of the pes (figure 4).

Figure 1), A) *Lambeosaurus* walking upright bipedally; B) walking horizontally bipedally; and C) walking quadrapedally.

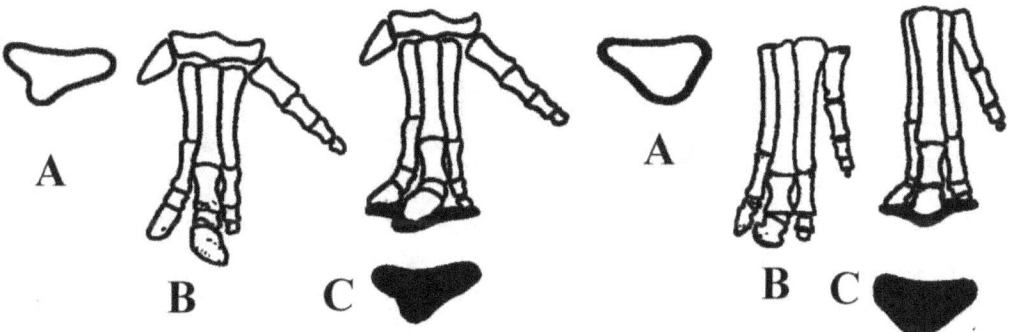

Figure 2), A) Iguanodontid manus print; B) manus; and C) *Iguandon* manus, showing how the manus into the print.

Figure 3), A) Hadrosaur manus print; B) manus: and C) Hadrosaur manus, showing how the manus fits into the print

The iguandontid tracks also show that these dinosaurs occasionally walked on all fours. Track Ca has 10 less manus prints than on Cb has only 1 manus print. This may be due to preservational bias, or that the print layer isn't the top of the tracklayer which Wright finds highly unlikely. Her conclusion is that they walking only occasionally quadrapedally.

This new interpretation seems strange, but it is supported by both footprints and the skeleton. What it certainly indicates is that not all ornithopods walked the same way and the trackways and trackway makers need to be looked at anew.

Bibliography

Currie, P. J., Nadon, G. C., and Lockley, M. G., 1991, Dinosaur footprints with skin impression from the Cretaceous of Alberta and Canada: Canadian Journal of Earth Sciences, v. 28, p. 102-115.

Ford, T. L., 1997, How to Draw Dinosaurs. *Iguanodon*: Prehistoric Times, n. 24, p. 30.

Wright, J. L., 1998, Ichnological evidence for the use of the forelimb in Iguanodontid locomotion: Special Papers in Paleontology, n. 60, p. 209-219.

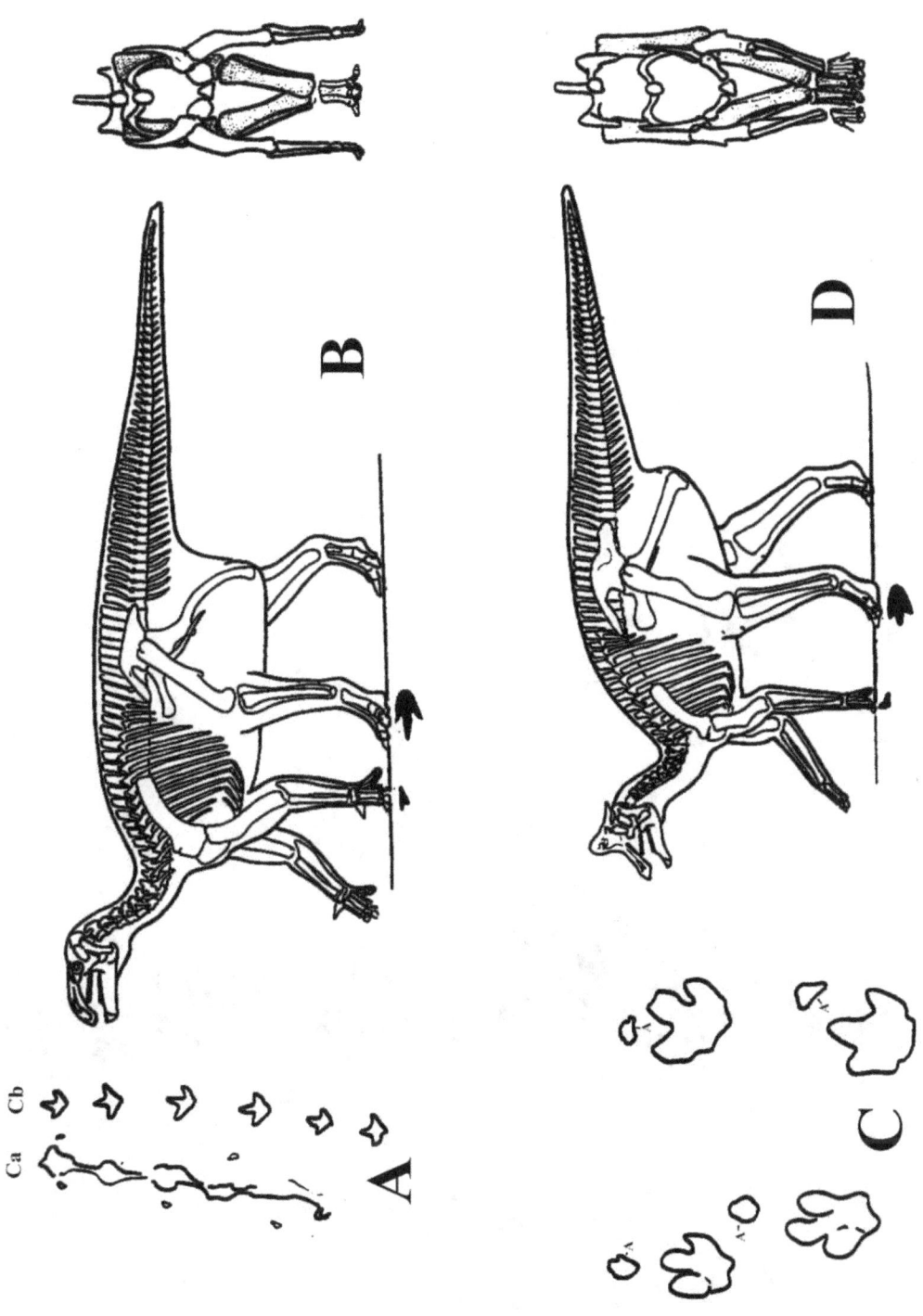

Figure 4), A) *Iguandon* trackway; B) side view of skeleton showing how the trackways were made, and C) cross-section of the skeleton; D) *Caririchnium* trackway; E) side view of skeleton showing how the tracks were made, and F) cross-section of the skeleton.

For the Dinosaur Collector and Enthusiast

PREHISTORIC TIMES

Dec/Jan 2000   No. 39

A Visit To DinoLand
Dino Model Kit
& Book Reviews
& Much More!

THE PT INTERVIEW:
Paleontologist John Horner
& Kaiyodo Sculptor
Kazunari Araki

U.S. $5.95 • Canada $6.95

0 74470 92325 1   12>

Ford, T. L., 1999, How to Draw Dinosaurs. Armor in 1 or 2 rows? Prehistoric Times, n. 36, p. 14.

Chapter 16

Armor in 1 or 2 rows?

The 'primitive' characteristic for dermal armor (osteoderms, scutes, spikes, or plates) is to have two rows down the midline of the back. This is true for thecodonts (or pseudosuchians), crocodilians, and dinosaurs to name a few. Stegosaurs are the most famous for having two rows of plates down the midline, with only *Stegosaurus* itself having offset plates from over the hips to the nearly the tip of the tail. Ankylosaurs have the most armor of any known group of dinosaurs. I won't be getting heavily into ankylosaurs but will make a few comments on new interpretations and new genera.

One of the things that bugs me is to see an illustration of *Scutellosaurus* with only one row of armor down its back. Ok, it's not a real famous dinosaur, but still... In Colbert's monograph he specifically said there are left and right small, laterally compressed posterior pointing plates (figure 1). Also, from these scutes are several other lateral rows of scutes down the body. The fault just may lie with the monograph itself. The main problem is that illustrations of *Scutellosaurus* are in lateral view, and it shows only one row of scutes down the back. This presents a problem that paleontologist need to address when they publish articles. They need to make it clear just how they interpret an animal and to illustrate it so the public, illustrators understand it.

Also, *Scelidosaurus* had two rows of plates down the mid-line of the back (figure 2) and several down the sides of the body. So, when illustrating, sculpting one of these animals, you should show two rows down the back.

Now onto *Sauropleta*. *Sauropleta* is one of the best-known nodosaurs and thanks to Ken Carpenter, one that shows how the armor was arranged on the animal. In his paper in 1984 (which is the main study for the animal) he illustrated the animal with one row of three large posteriorly pointing scutes down the side of the neck. Down the midline of the neck was small oval scutes. Behind the last scute was a longer larger scute.

There is a very dramatic change to *Sauropelta*, thanks to Ken Carpenter and Jim Kirkland (1998). They show another row of scutes beneath the side row of neck scutes (figure 3). While looking over the collection of *Sauropelta* material at the American Museum of Natural History, Ken noticed that AMNH 3035 had the cervical scutes in articulation, as did the Yale Peabody Museum specimen, YPM 5178. Also Carpenter and Kirkland believe that the lateral scutes down the tail were actually larger. This gives *Sauropelta* a very spike look to it.

The armor in ankylosaurs varies greatly from polocanthids, ankylosaurinae, nodosaurinae, and shasmosaurinae. For now I'll refrain from illustrating these subfamilies/families, and am hoping to study these groups in detail.

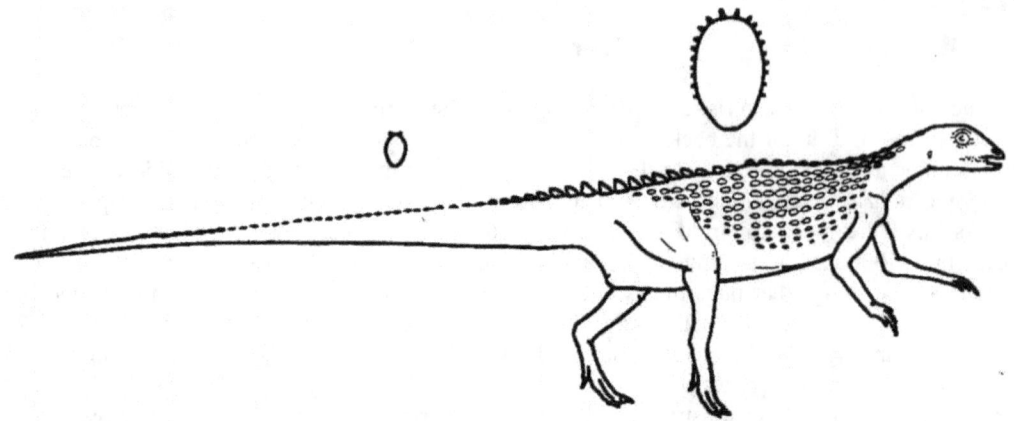

Figure 1. *Scutellosaurus* in lateral view and cross-section of the middle of the body and tail.

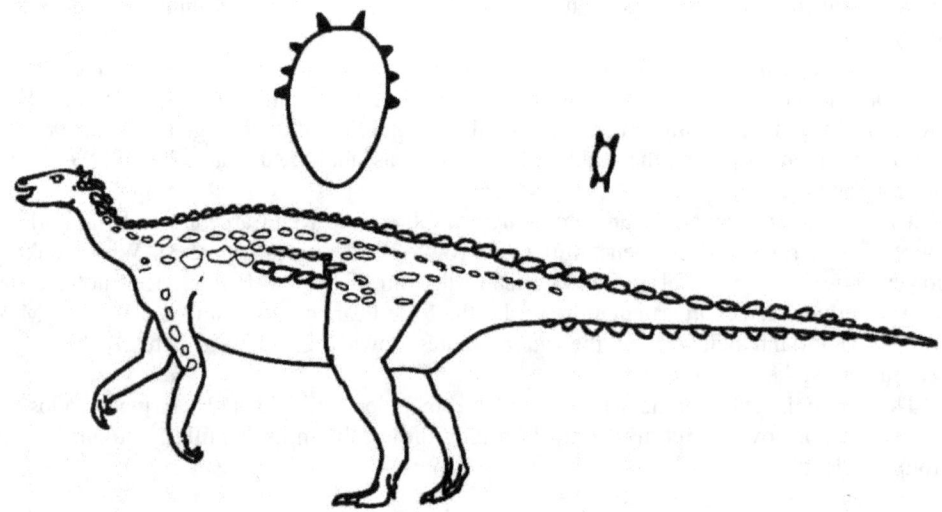

Figure 2. *Scelidosaurus* in lateral view and cross-section of the middle of the body and tail.

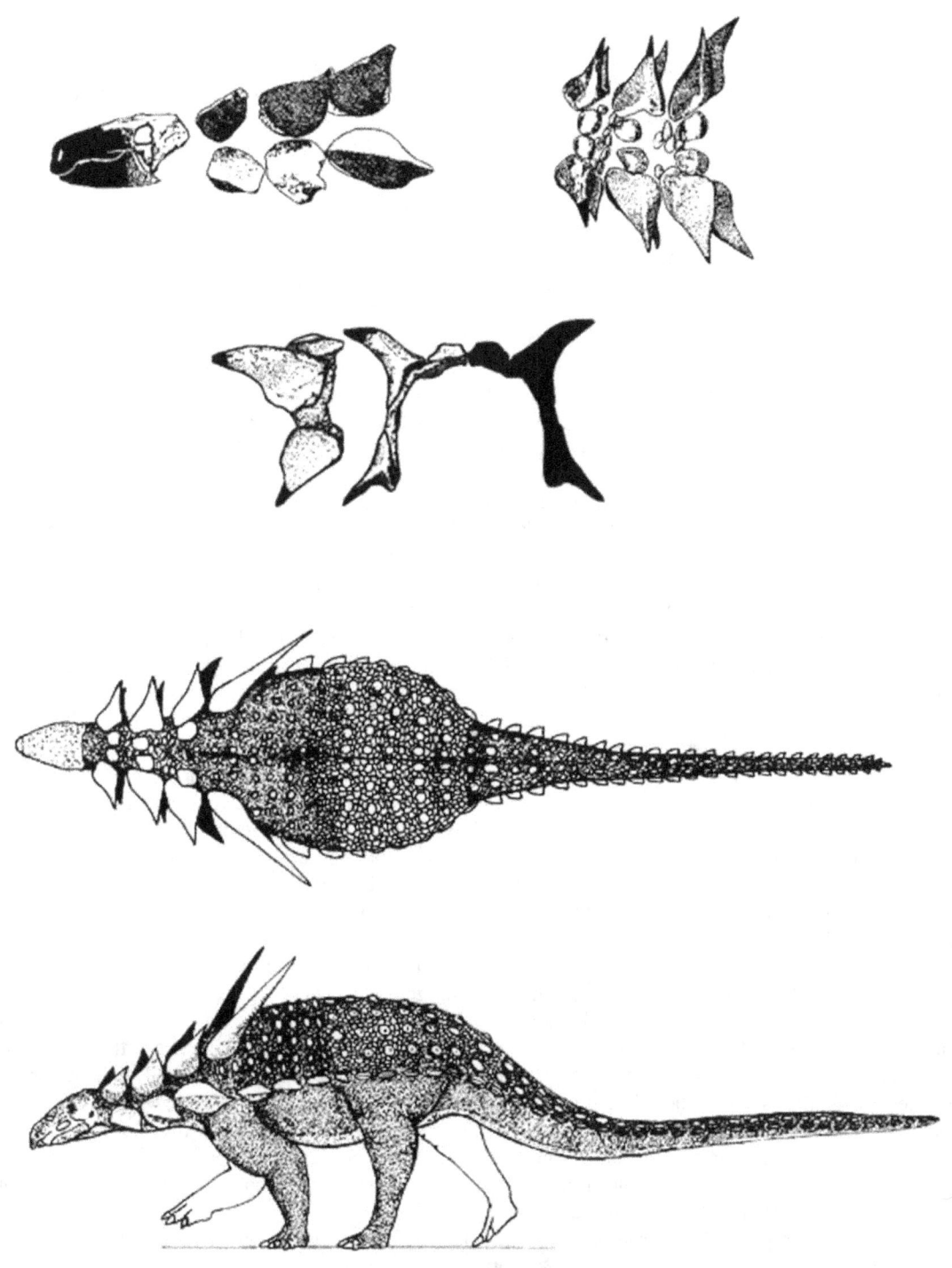

Figure 3. *Sauropleta* showing the new arrangement of cervical armor. Side and top view of AMNH 3035, anterior and side view of YPM 5178.

113

More Ankylosaurs

Several new ankylosaurids have been recently described. In the now famous Nature issue with *Caudipteryx* is a description of a new Morrison Ankylosaurid, *Gargoyleosaurus sparkpini* Carpenter, Miles & Cloward, 1998 (figure 4 a, housed at the Denver Museum of Natural History) are from the famous Bone Cabin Quarry of Albany County, Wyoming. This quarry has been producing dinosaurs for more than 100 years, *Apatosaurus, Brachiosaurus, Camarasaurus, Diplodocus, Allosaurus, Ornitholestes, Dryosaurus* and *Stegosaurus*. This new ankylosaur shows both nodosaur and ankylosaur characters. From the top, the skull resembles many of the late Cretaceous ankylosaurs. The skull is triangular in shape and there is the presence of large, triangular scutes at the rear corners (thus the name 'gargoyle reptile', in reference to the gargoyle-like appearance of the skull in profile). The scooped beak is similar to nodosaurs. Also known are skeletal elements as well as armor. There are two rings of neck armor (like ankylosaurs, nodosaurs have this ring split in two down the middle) with the first ring having 5 partly fused scutes and the second having 6 partly fused scutes. And at least two elongated spines extended from both shoulders, which is a nodosaur characteristic. This specimen will be very important in showing how ankylosaurs evolved.

A new ankylosaur from New Mexico shows similarities to Asian ankylosaurs. *Nodocephalosaurus kirklandensis* (Sullivan, 1999), (figure 4 b, housed at The State Museum of Pennsylvania, Harrisburg), was found west of Willow Wash, San Juan County, New Mexico and is from the Kirkland Formation, Campanian, Late Cretaceous. The skull is incomplete, known mainly from the left side and is slightly crushed. Even though the snout is directed downward, the crushing didn't turn the snout downward and the skull had an arch to it, not seen in any other ankylosaur. What is known shows similarities to *Saichania* and *Tarchia* in that the dermal armor on the skull is semi inflated to bulbous, polygonal, and bilaterally and symmetrical arranged on the top of the skull. Also the spikes on the back of the skull are more like the Asian forms than North America forms. What this suggests/supports is that some time in the Late Cretaceous, North America and Asia were connected.

The long awaited description of *Gastonia* is out. *Gastonia burgi* Kirkland, 1998 (figure 5, housed at the College of Eastern Utah with referred material at Brigham Young University). The holotype material is from Gaston Quarry, Cedar Mountain Formation, Early Cretaceous and consists of a skull and skeletal remains. *Gastonia* is a North American polacanthid. The skull is triangular in top view and has a little or no osteoderms on the top. Two things about the skull are (so far) unique to *Gastonia*. The orbits are nearly complete and facing forward giving the animal stereoscopic vision (more so than in *Tyrannosaurus*) and the front of the jaws is concave. Unfortunately no lower jaw is known. What Kirtland thinks may be the reason for the concavity is to allow the tongue to stick out there. One of the really interesting things about *Gastonia* is the armor. The first two scutes on the neck are small and are triangular in dorsal views. The next 3 scutes are very large with the first being the largest. Each one of these scutes has a groove on the back end that the proceedings scute fits into, thereby as the animal turned his neck the scutes acted like shears. The following two scutes are similar in shape, but without the groove. The lateral armor on the pelvis is much smaller. The lateral tail scutes were large and compressed top to bottom. They are triangular in shape with the plates pointing posteriorly. The tail is long. The scutes down the middle of the body stick straight up. The first two are small, with the next one more pointed and tall, the third one is very large and turns slightly outward. The next two are smaller and the last one is just a bump. In between the midline and lateral scutes were one, possibly two rows of scutes. The pelvis has a 'shield' covered with oval scutes separated by smaller scutes. The tail had a row of tear dropped scutes. *Gastonia* shows similarities to a Morrison polacnathid named *Mymoorapelta*. What these polacantids indicates is that there was a land bridge from North American and Europe during the Late Jurassic and Early Cretaceous. The material from Dalton Wells is at BYU. The new material also consists of some skulls (no lower jaws still), with one belonging to a juvenile and other skeletal elements.

In the same paper that Carpenter and Kirkland, 1998, mention a new skull from Carbon County Utah, (figure 4c). The skull is from the Cedar Mountain Formation, Early Cretaceous (and will be described in detail in a forthcoming paper by the authors) (**Editors note: It was named *Cedarpelta bilbeyhallorum* Carpenter, Kirkland, Burge & Bird, 2001**) The skull has a strong resemblance to Shamosaurid ankylosaurs. Shamosaurs have smooth or nearly smooth skulls with slightly pointed box scutes on the end of the skull. The new skull is fragmentary, but shows that the top of the skull was smooth and had box like scutes on the edge of the skull. The animal would have been larger than *Ankylosaurus*. Also in said paper, they show that the 'club' on the end of *Polacanthus* is actually fused tendons, tail vertebrae and lateral scutes from around the middle (?) of the tail. What these ankylosaurs seem to prove is the existence of land bridges to both Europe and Asia during the Late Mesozoic.

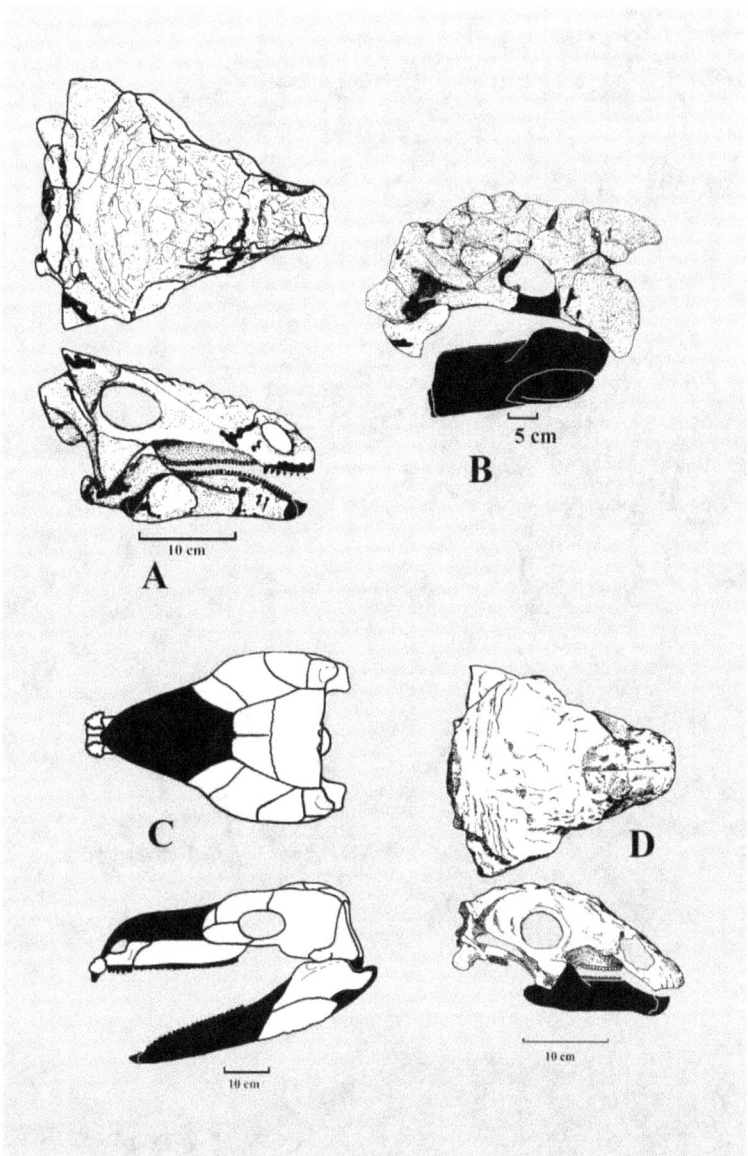

Figure 4) A) *Gorgoyleosaurus parkpini* DMNH 27726, after Carpenter, et al., (1998); B) *Nodocephalosaurus kirtlandensis* SMP VP-900, after Sullivan, (1999); C) *Cedarpelta bilbeyhallorum* Carpenter, et al., 2001; D) *Gastonia burgei*, CEUM 1307, modified after Kirkland, (1998)

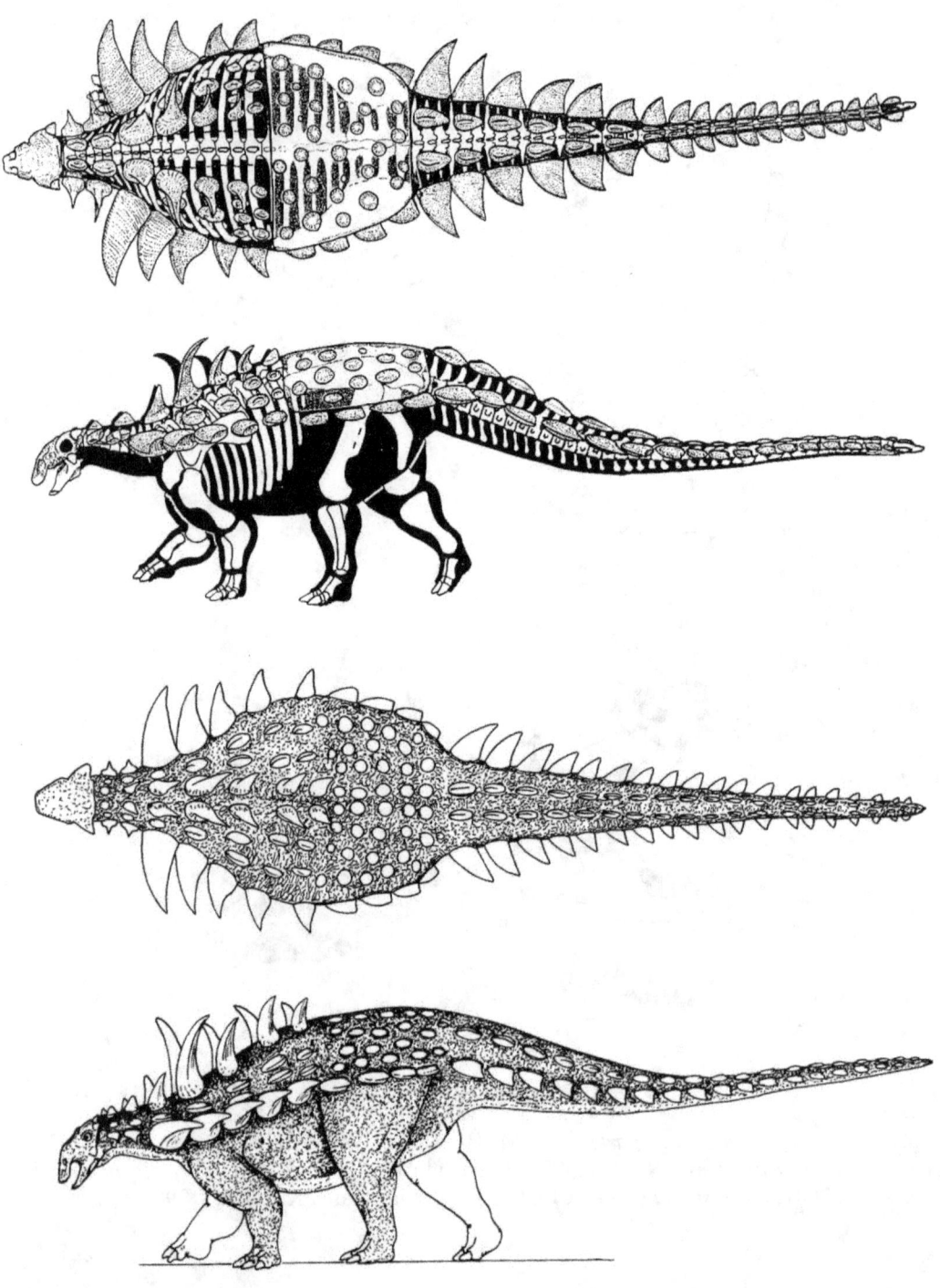

Gastonia burgei, CEUM 1307, modified after Kirkland, (1998); skeleton and life reconstruction.

Bibliography:

Carpenter, K. 1984. Skeletal reconstruction and life restoration of *Sauropelta* (Ankylosauria: Nodosauridae) from the Cretaceous of North America. Can. J. Earth Sci. Vol. 21: 1491-1498.

Carpenter, K., Kirkland, J. I., Burge, D., and Bird, J., 2001, Disarticulated skull of a new primitive ankylosaurid from the Lower Cretaceous of Utah: In: The Armored Dinosaurs, edited by Carpenter, K., Indiana University Press, p. 211-238.

Carpenter, K., and Kirkland, J. I., 1998, Review of Lower and Middle Cretaceous Ankylosaurs From North America. In: Lower and Middle Cretaceous Terrestrial Ecosystems. Edited by Spencer G. Lucas, James I. Kirkland and John W. Estep, New Mexico Museum of Natural History and Science, Bulletin 14: 249-270.

Colbert, E. H. 1981. A Primitive Ornithischian Dinosaur from the Kayaenta Formation of Arizona. Museum of Northern Arizona Press, Bulletin Series 53: 1-61.

Kirkland, J. I., 1998, A polacanthinae ankylosaur (Ornithischia: Dinosauria) from the Early Cretaceous (Barremian) of Eastern Utah: In: Lower and Middle Cretaceous Terrestrial Ecosystems, edited by Lucas, S. G., Kirkland, J. I., and Estep, J. W., New Mexico Museum of Natural History and Science, Bulletin n. 14, p. 271-281.

Sullivan, R. M., 1999, *Nodocephalosaurus kirtlandensis*, gen. et. sp. nov., a new Ankylosaurid Dinosaur (Ornithischia: Ankylosauria) from the Upper Cretaceous Kirtland Formation (Upper Campanian), San Juan Basin, New Mexico: Journal of Vertebrate Paleontology, v. 19, n. 1, p. 126-139.

For the Dinosaur Collector and Enthusiast

PREHISTORIC TIMES

October/November 1999   No. 38

The PT Interview: Ely Kish

Disneyworld's New Dinoland

BBC's "Walking With Dinosaurs"

U.S. $5.95 • Canada $6.95

Ford, T. L., 1999, How to Draw Dinosaurs. The Eye!!!: Prehistoric Times, n. 37, p. 14-15.

Chapter 17

The EYE!!!

Recently Mike Fredricks asked me about the eyes of dinosaurs. Could dinosaurs have had slit pupils like cats and crocodiles rather than round pupils, etc. I really had not thought much about it at the time. I had always simply drawn them with round pupils.

In 1979, John McLoughlin published a book called Archosauria, A New Look at the Old Dinosaur. In his book, McLoughlin drew many of the ornithischia with rectangular pupils. I never liked this because it looked too much like a mammal to me, but what if I'm wrong and McLoughlin was right? Well, I had to do more research than usual and this article will be more of a biology lesson, but it will hopefully open the eyes (so to speak) of the artist to more options. Thank you Mike for requesting this.

The natural state of the eye is to be round. The external and internal forces naturally make the eye this shape. The sclerotic rings can aid in retaining a different shape for the eye, other than round, but the sclerotic ring has nothing to do with pupil shape, so no further discussion about the sclerotic ring is needed. What really needs to be discussed is the function of the shape of the pupil in animals. There are two major types of pupils, round and slit (or elliptical). The second type can vary greatly, even within species. A round pupil has no more say in whether or not the animal is nocturnal or diurnal. It is true that a slit pupil is more sensitive to light than a circle and many nocturnal animals have slip pupils, but Owl's have round pupils and they are mainly nocturnal. An oval pupil can contract only so far, where as a slit pupil can contract either completely or nearly so into a straight line (some animals will have a slight oval at each end).

In amphibians the vast majority of pupils are slit, either horizontal or vertical. They can vary from rounded edges, to diamond shape (one frog has heart shaped pupils). The pupils of reptiles have larger variety of shapes. There are many with circular pupils but the majority are slit, with some being vertical ellipses or vertical ovals. The diurnal lizards have circular pupils, while the nocturnal have slits (most of the snakes, lizards, Geckos [with many have very strange slits], night lizards, the *Sphenodon* and the crocodilians).

The shape of the eyes in mammals vary greatly. They can be circular, with the majority being the vertical slit, the vertical oval, the vertical ellipse, the horizontal oval, the horizontal slit or the horizontal rectangle. Ungulates have a horizontal slit or horizontal rectangle. The majority of ungulates have the horizontal slit or the horizontal rectangle; also those with brow ridges, horns, etc. (i.e. horses, cattle, deer, sheep, goats, hippopotamus, camels, giraffes, and lamas). All of these animals, to some degree, are open planes animals and need to keep an 'eye' out for carnivores. Only the elephant (with its large size) has circular eyes and may not need to keep a look out for carnivores. Birds on the other hand, are all nearly circular pupils, with only a few diverging from this. So, all nocturnal birds have circular eyes. The King Penguin contracts it pupils in a series of polygonal shapes, to a square. The tiger bittern has a horizontal elongate pupil. And only one bird has a vertical slit, the black skimmer. (figure 1).

The function of the two types of pupils vary. A round eye is less sensitive to light, and the size of the pupil will help in the clarity of the image (along with other cornea, eye shape, etc.). The slit eye is more light sensitive, but neither is really needed to be a nocturnal or diurnal, as stated before. Being a carnivore does not necessarily require a certain pupil type. For example, a dog has a round pupil, and a cat has a slit pupil. In mammals that are range animals, slit eyes help extend the field of vision without increasing the size of the eye, and as stated before, helps the animal keep a look out for perdition.

Then what kind of pupils do dinosaurs have? This is unknown and probably will never be known. Theropods, at least the avian theropods (birds) and those more closely related to birds, I believe would have the circular pupils like birds (dromaeosaurs, troodontids, oviraptors, etc), while the others have had circular as well as slit pupils (allosaurs, megalosaurs, abelisaurids, ceratosaurids, etc). After reading up on then pupil's shapes, I believe that the herbivores, at least the 'range' or herding ones and the ones with brow horns, etc., had horizontal slits to help them keep an eye out for predators (as McLoughin illustrated.) (figure 2).

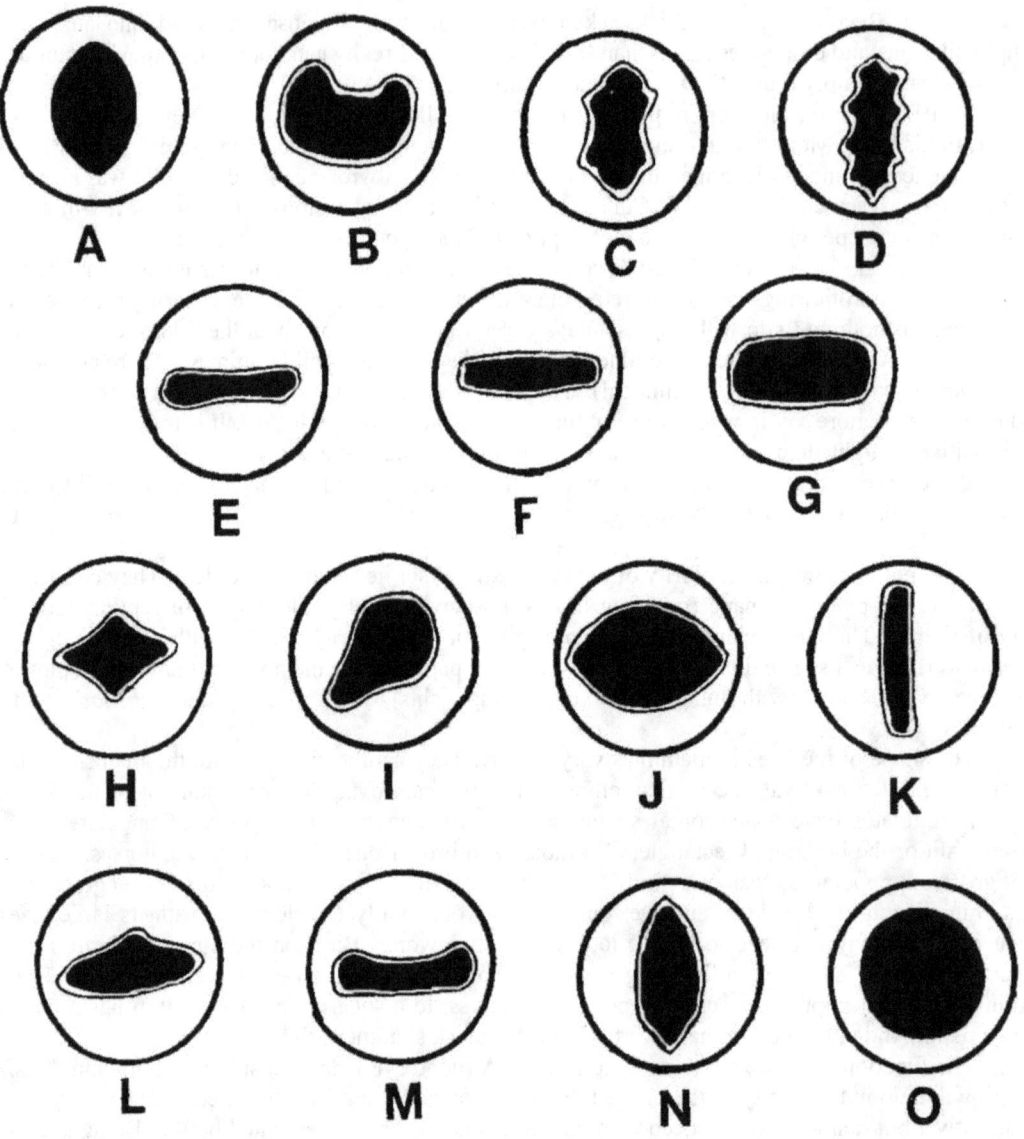

Figure 1). Pupils of animals. A) Iguana; B) Dolphin; C and D) Geckos; E) Whale in contraction; F) Deer in contraction; G) Deer normal; H, I and J) Frogs; K) Python in contraction; L); Hippopotamus in contraction; M) Barbary Sheep in contraction; O) Crocodile and P) Bird.

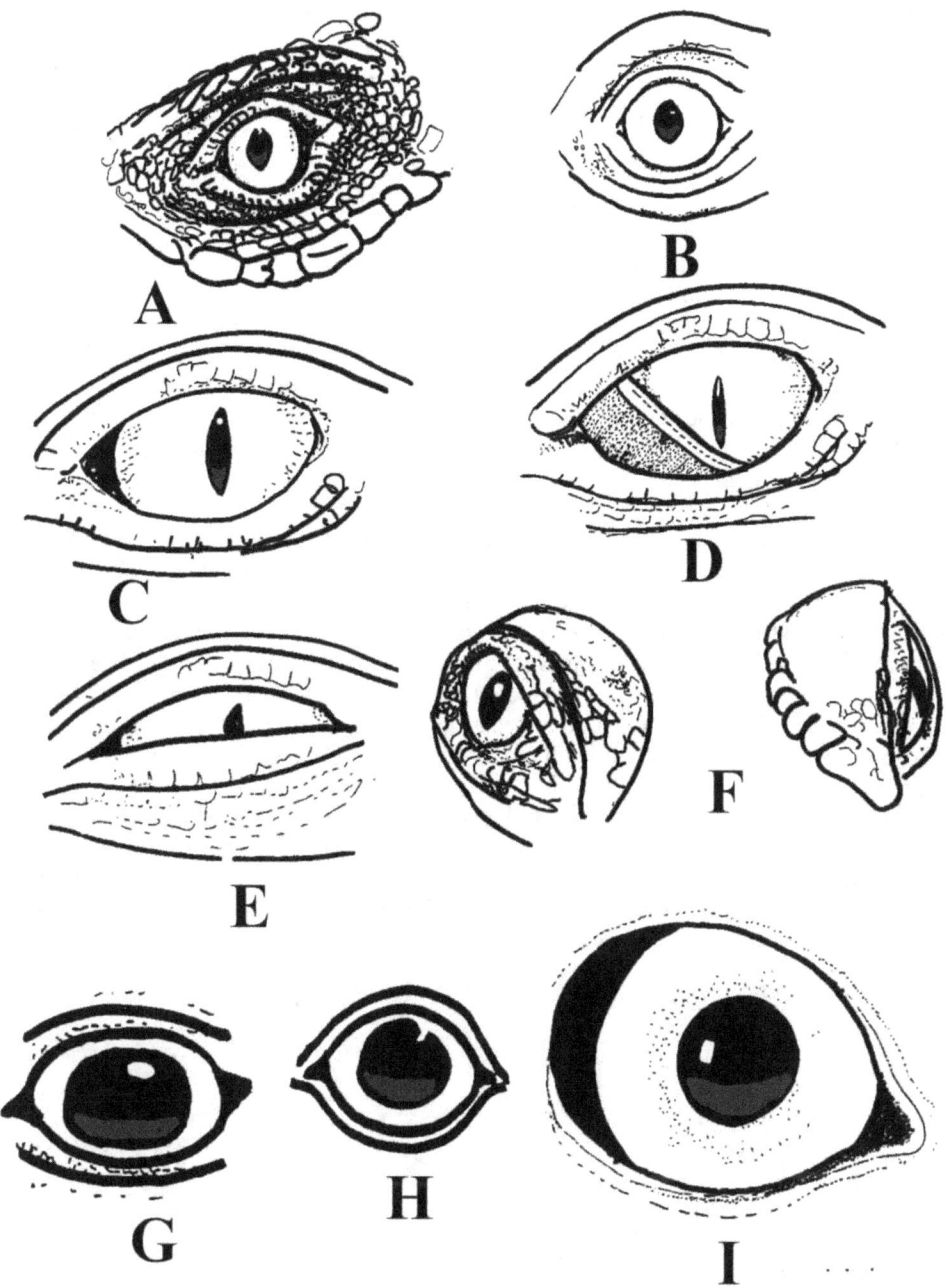

igure 2). Eyelids of animals. A and B) Iguana's; C) Crocodile; D) Crocodile showing nictitating membrane closing; E) Crocodile closing it's lower eyelid; F) Three quarters front view of a crocodile; G) Bird; H) Dolphin and I) Hawk.

What about the eyelids? The nictitating membrane? The nictitating membrane is a type of eyelid, a single 'sheet' of skin that can in some cases, be nearly transparent. The need for eyelids may have started when animals crawled out of the water and onto dry land. The eyelids help shield the eyes from glare, dust, running into plants, burrowing, to help keep the eye from dehydrating, and lastly to voluntarily to block out light, i.e. to sleep or rest. All the totally aquatic animals lack eyelids, (with only a few fish, having what might be called rudimentary eyelids.) This would mean that the totally aquatic reptiles, ichthyosaurs and plesiosaurs, lacked eyelids of any kind. In whales and dolphins, the eyelid is nearly flush with the head, because they don't need to 'shield' the eye from glare or stop the eye from dehydrating, getting dust, etc into it. Frogs and toads have an active nictitating membrane, which moves more than the eyelid, while in salamanders, the nictitating membrane is rudimentary. The lizards that lack nictitating membrane are nocturnal, geckos, chameleons and Australian skinks. The Australian skinks and geckos actually have had the lower eyelid attached to the upper eyelid, like snakes and form what is called a brille. This has turned into a transparent non-moveable eyelid, which protects the eye.

The Sphenodon has a nictitating membrane. Crocodiles have a nictitating membrane which they use to cover the eye as they swim threw the water. The same is true for turtles. Birds and the majority of mammals also have a nictitating membrane. Knowledge of the nictitating membrane is practically useless for the artist, but is an interesting part of the animal. In birds the nictitating membrane helps lubricate the eye and protect it, and when birds sleep, only then do they close the eyelids.

The artist can use the closing of the eyelid in his work. For the most part, all lizards, crocodilians and some birds, only close the eye via the lower eyelid. This means the lower eyelid closes up to the upper eyelid. Some birds use both upper and lower eyelids. In mammals, it is usually the upper eyelid that is more mobile. (Figure 3).

One more thing the artist can use, and that is how to depict the eyelids. Lizard's eyelid has a major upper 'eyelid', and several smaller 'eyelids', thus forming a circle with horizontal points at each end where the eyelids attach. Some birds; hawks for example, have only one side of the eye with a 'point' and a rounded posterior end. The illustrator, if they'd like, can use a hawk's eyelid style for theropods that are more closely related to birds (i.e. dromaeosaurids, oviraptors, troodontids, etc) and a lizard eye could be used for all other dinosaurs. Crocodilian eyelids could be used for, obviously for crocodilians; but also for all other aquatic reptiles with eyes on top of their heads, i.e. phytosaurs, protochampsids, etc.

I hope this will help the illustrator with more options to use in their work, (I know it will in mine).

Biblography

McLoughlin, J. C., 1979, Archosauria, A New Look at the Old Dinosaur: The Viking Press, 117pp.

Prince, J. H., 1956, Comparative Anatomy of the Eye: Charles C. Thomas publishers: Various chapters used.

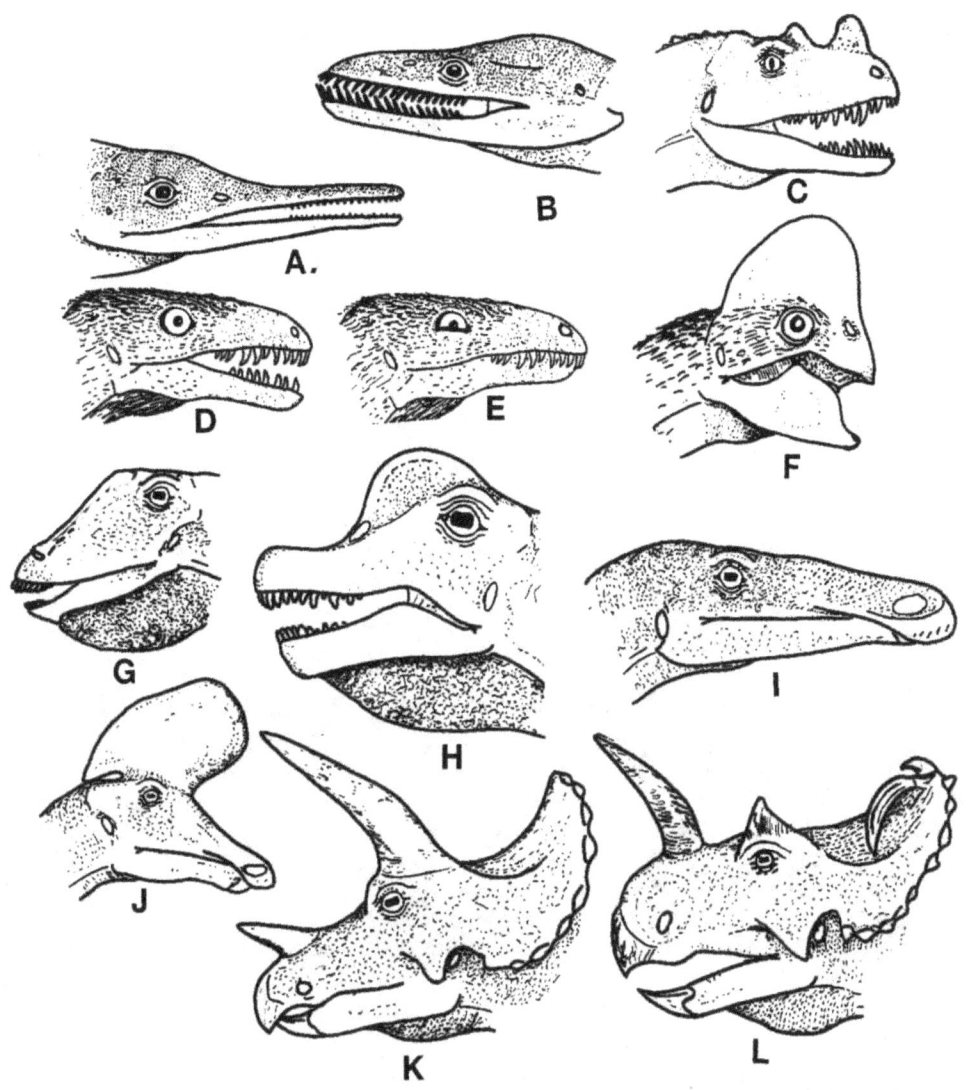

Figure 3. Illustrations showing possible pupils and eyelids of prehistoric animals; A) The Ichthyosaur *Ophthalmosaurus*, with Dolphin like eye; B) *Muraenosaurus* with Dolphin like eye; C) *Ceratosaurus* with lizard like eye; D) *Dromaeosaurus* with hawk like eye; E) *Dromaeosaurus* closing it's lower eyelid; F) *Oviraptor* with hawk like eye; G-L different plant eating dinosaurs with ungulate like eyes; G) *Diplodocus*; H) *Brachiosaurus*; I) *Anatotitan*; J) *Lambeosaurus*; K) *Triceratops* and L) *Centrosaurus*.

For the Dinosaur Collector and Enthusiast

PREHISTORIC TIMES

June/July 1999   No. 36

Continuing Our
PT Interview:
Gregory S. Paul

Also:
Paleontologist
Mark Norell

Dinosaur Modeling Tips & Reviews!

U.S. $5.95 • Canada $6.95

Ford, T. L., 1999, How to Draw Dinosaurs. Roman noses and Cassowary crests: Ornithopoda heads. Part 1; *Tenotosaurus*, *Muttaburrasaurus*, *Altirhinus* and hadrosaurian hadrosaurs: Prehistoric Times, n. 38, p. 14-15.

Chapter 18

Roman noses and Cassowary crests: Ornithopoda heads. Part 1; *Tenotosaurus*, *Muttaburrasaurus*, *Altirhinus* and hadrosaurian hadrosaurs.

The crested hadrosaurs are one of the most recognizable of all ornithopods. Hadrosaurs are grouped into two main groups, the hadrosauridae or the flat headed hadrosaurs, and the lambeosauridae, the helmet crested hadrosaurids. The helmet crested lambeosaurs are a staple in what dinosaurs look like to eye of the general public. New finds are shedding light on lambeosaurs, showing that the earliest ones lacked crests (See Prehistoric Times number 35, page 15). Some flat headed hadrosaurs have a spike, or bump on the nose or over the eyes with some earlier ornithopods having a bump on the nose. These can be grouped by morphs. Why the ornamental head crests? Possibly for individual identification. At the Page Museum (the museum at the Tar Pits of Rancho La Brea in Los Angles), there is a wall exhibit that shows nearly a hundred different skulls of the Dire Wolf. This exhibit shows how widely the appearance of the skull varied. This is due to sexual dimorphism (or the difference between males and females), individual variation or identification in a herd, and used for attracting males and females to one another.

For the most part, all the other ornithopods have flat or nearly flat heads. Hypsilophodons lack any cranial crest or ornamentation of any kind. *Tenontosaurus*, the famous meal for *Deninonychus* has a fairly large nose. *T. tillettorum* is the cow of the Early Cretaceous. It is known from Montana, and Wyoming. (figure 1a). The skull of *T. tillettorum* was intact, but then dropped and later glued together. There is a ridge on the nose giving the animal a small bump or a roman nose (not related to the hadrosaurs roman nose).

There are several types of *Tenontosaurus*, and the skulls vary from region to region. *T. dosii,* is from Doss Ranch, Parker County, Central Texas, Twin Mountains Formation, Early Cretaceous (figure 1b) appears to have lacked a bump, but the skull is crushed and it is difficult to tell how the top of the skull originally appeared. There is a *Tenontosaurus* sp from Oklahoma that appears to have lacked the bump, but has a higher nose. This may be a new species of *Tenontosaurus* (figure 1 c). *Tenontosaurus* is also known from Montana, Wyoming, Utah, Arizona, Idaho, Texas and Oklahoma.

Exactly what *Tenontosaurus* is, is very debatable. Paleontologists have questioned whether it is a hypsilophodontid or not. *Muttuburrasaurus*, the best known Australian ornithopod (which is very difficult to place within the ornithopoda) has a large roman nose. The type species *M. langdoni* has an incomplete skull so determining exactly how the head looked is debatable (figure 1d.). A new specimen has a different kind of roman nose and seams to indicate a different species. (figure 1e.).

Nearly all the iguanodontids have a flat head with *Altirhinus* having the largest nose. It is the nose itself that is inflated and not a large flat crest like that of *Gryposaurs*. (figure 1f). *Altirhinus* was referred to *Iguanodon orientalis* for years, yet it wasn't accurately studied until David Norman worked on the specimen for several years in the early 1990's. He concluded that there is an *Iguanodon orientalis*, but it isn't *Altirhinus*. *Altirhinus kurzanovi* is from Khuren Dukh, Mongolia, Early Cretaceous, in age.

One of the most famous iguanodons with the most distinctive head is *Ouranosaurus* (figure 1g). *Ouranosaurus nigeriensis* was found 7 km south-east of El Rhaz, Gadoufaouna, Niger, Egypt, and comes from the El Rhaz Formation, Early Cretaceous. The skull is long with a flat nose, but has two small bumps just in front of the eyes. *Ouranosaurus* is most famous for its high neural spines. It has been suggested that *Iguanodon* and *Ouranosaurus* gave rise to hadrosaurids and lambeosaurids (respectively).

Hadrosaurid hadrosaurs have a relatively small cranial crests and these crests types can be divided in to several morphs. Gryposaurs have a Classic 'Roman' nose. *Gryposaurus notobilis* (which had been referred to *Kritosaurus* for several years) has the classic 'roman' nose (figure 2a, b). *Gryposaurus* is from Dinosaur Provincial Park in Alberta Canada. *Gryposaurus latidens* is known from the Two Medicine River, Pondera County, Montana and is from the Lower Two Medicine Formation, Late Cretaceous. This *Gryposaurus* has a smaller more pointed roman nose (figure 2c, d). The type *Kritosaurus, K. navajovius* comes from the Late Cretaceous of New Mexico (figure 2d). It lacks the front end of the skull, so the

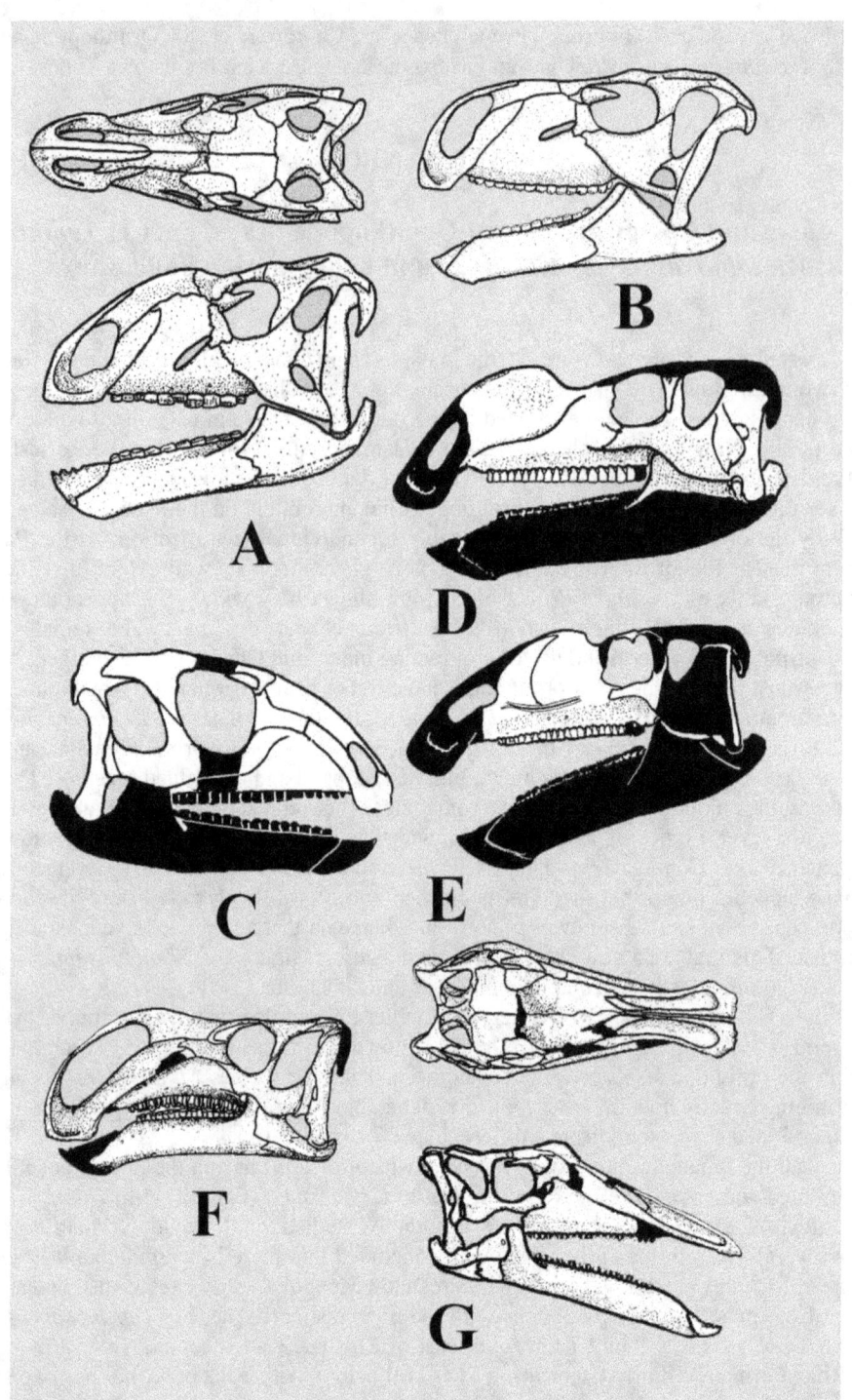

Figure 1). Early Ornithopods; A) *Tenontosaurus tillettorum*, top and side view; B) *Tenotosaurus dosii*; C) *Tenotosaurus* sp from Oklahoma; D) *Muttuburrasaurus langdoni*; E) *Muttuburrasaurus* sp; F) *Altirhinus kurzanovi*; G) *Ouranosaurus nigeriensis*, top and side view.

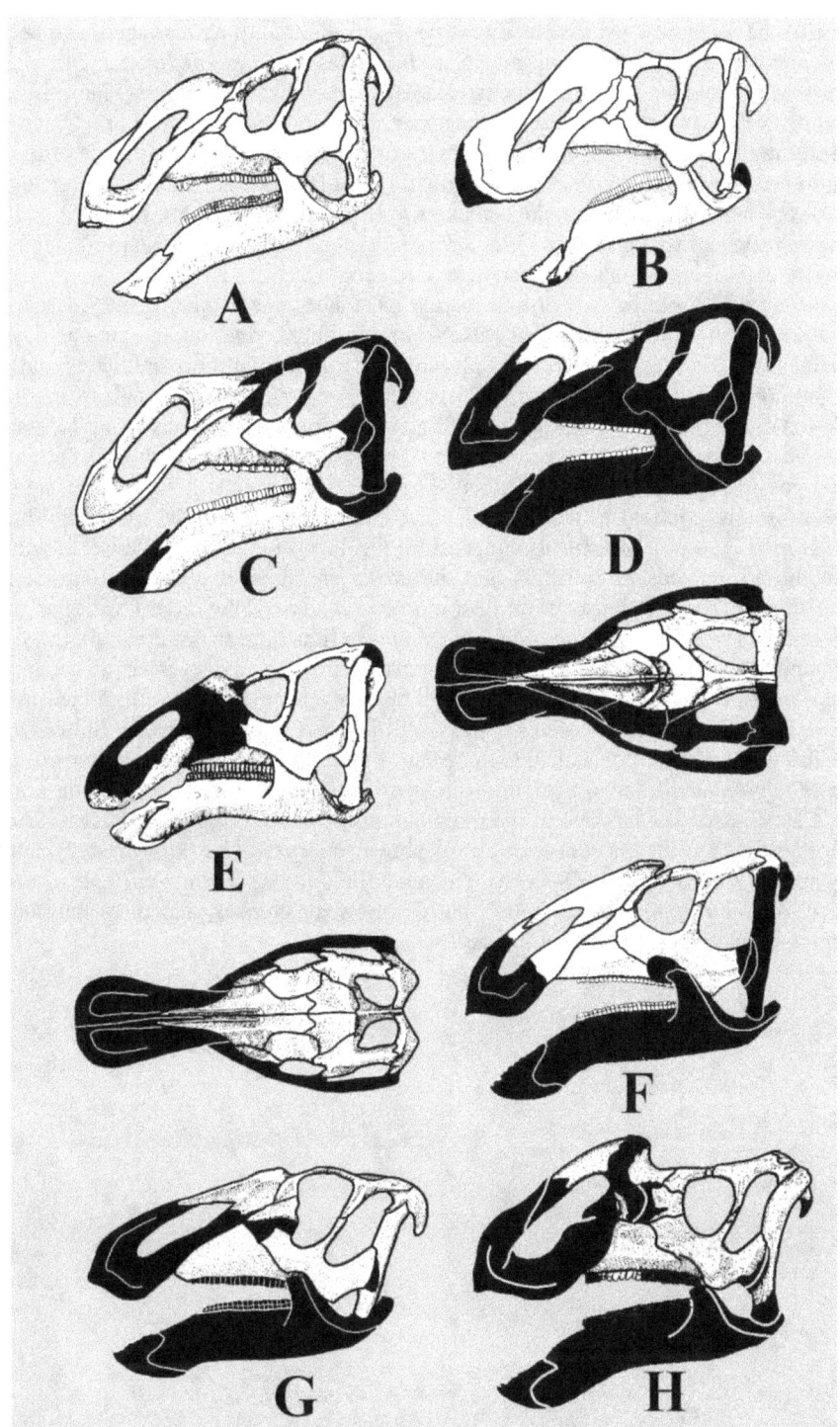

Figure 2). Gryposaurians Hadrosaurids ; A, B) *Gryposaurus notobilis* (Type specimen A, referred specimen B); C, D) *Gryposaurus latidens* (Type specimen C, referred specimen D); E) *Kritosaurus navajovius*; F) *Anasazisaurus horneri*, top and side view; G) *Naashoibitosaurus ostromi*, top and side view; H) *Aralosaurus tuberiferus*.

127

appearance of its nose is unknown. Two specimens that were referred to *Kritosaurus* have been changed to; *Anasazisaurus horneri* (Figure 2f), from the San Juan Basin, New Mexico, and is Early Kirkland Formation, Early Maastrichtian, Late Cretaceous; and *Naashoibitosaurus ostromi* (Figure 2g)**,** also from the San Juan Basin, New Mexico, but from the upper Kirkland Formation, Late Maastrichtian, Late Cretaceous. Both fragmentary and have different styles of nasal crests. The acceptance of these two genera is tentative by some. *Aralosaurus tuberiferus* is known from a fragmentary skull and fragmentary skeleton. The skull has a very large *Gryposaurus* like roman nose (figure 2h). This nose is higher and has a sharper crest. It comes from Beleutinskaia suite, 80 km north of Darmakchi (Dzhusaly Station), central Kazakhstan and is from the Beleutinskaya Formation, Early Late Cretaceous.

This next group could be called the maiasaurians. These hadrosaurs have a large more forward placed noses. *Brachylophosaurus canadensis* (also from the same formation and area as *Gryposaurus* and also from Montana) has a larger, more forward placed nose with a small horizontal crest at the back of the skull (Figure 3a). *Brachylophosaurus goodwini* has a smaller nose and comes from the Two Medicine Formation of Montana (Figure 3b) (**Editors note: Prieto-Marquez, 2005, has sunk *B. goodwini* into B. canadensis**). *Maiasaura peeblesorum*, one of the most famous and abundant hadrosaurs (from an estimated 1,000 to 10,000 possible specimens) has a *Brachylophosaurus* like nose (David Trexler had given a talk at an SVP that showed this, contrary to what Jack Horner had proposed) without the crest, and with a small bump over the eyes (figure 3c). *Maiasaura* comes from the Two Medicine Formation of Montana. One of the things that the artist needs to know, is that the crests or noses of these hadrosaurs, in fact on all hadrosaurs, are thin in cross-section or seen in top or dorsal view. They aren't wide, as has often been depicted. See the top view of *Brachylophosaurus canadensis* (figure 3a) to see how thin the skull is.

Saurolophinae hadrosaurids have a small crest in front of the eyes or a spike over the eyes. *Prosaurolophus maximus* has a long skull with a small bump over the eyes (also from the same horizon and area as *Gryposaurus*) (figure 4a). *Prosaurolophus blackfeetensis*, from Landslide Butte, Glacier County, Montana, Upper Two Medicine Formation, Campanian, Late Cretaceous, has a shorter skull with a wider bill (figure 4b). *Saurolophus osborni*, from the Red Deer River area, Horseshoe Canyon Formation, Early Maastrichtian, Late Cretaceous, has one of the most famous heads of the crested hadrosaurids (figure 4c). There is a short, pointed crest that curves upward above the eyes. The only crested hadrosaurid from Mongolia is *S. angustirostris* Nemegt, South Gobi, Mongolia, Nemegt Formation, Late Cretaceous (figure 4d). Both species of *Saurolophus* look similar, but do show differences, and there may actually be two different genera (speculated by George Olshevsky).

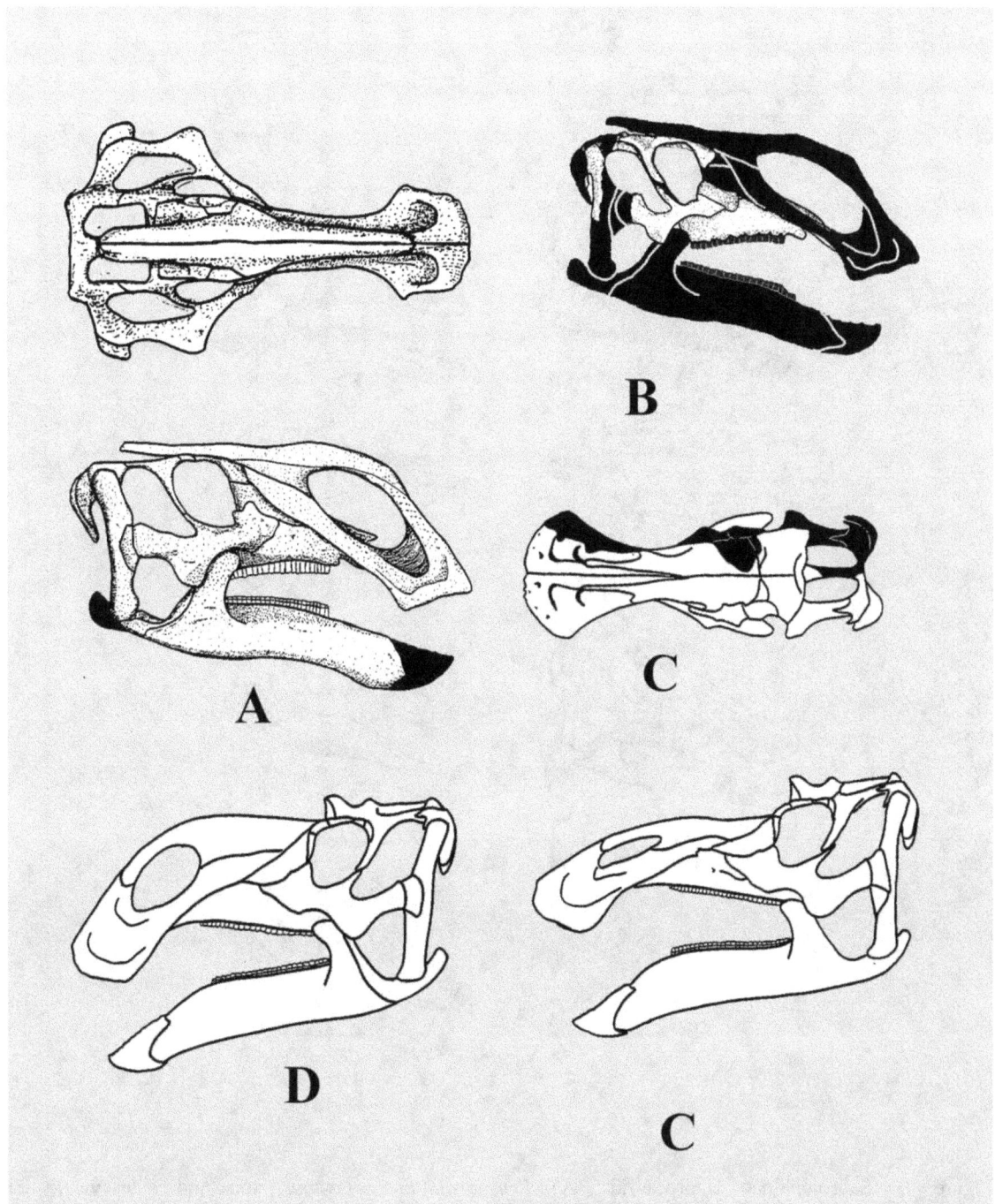

Figure 3). Maiasaurians Hadrosaurids; A) *Brachylophosaurus canadensis*, top and side view; B) *Brachylophosaurus goodwini*; C) *Maiasaura peeblesorum*, (right), top and side view after J. Horner, (left) view modified after David Trexler.

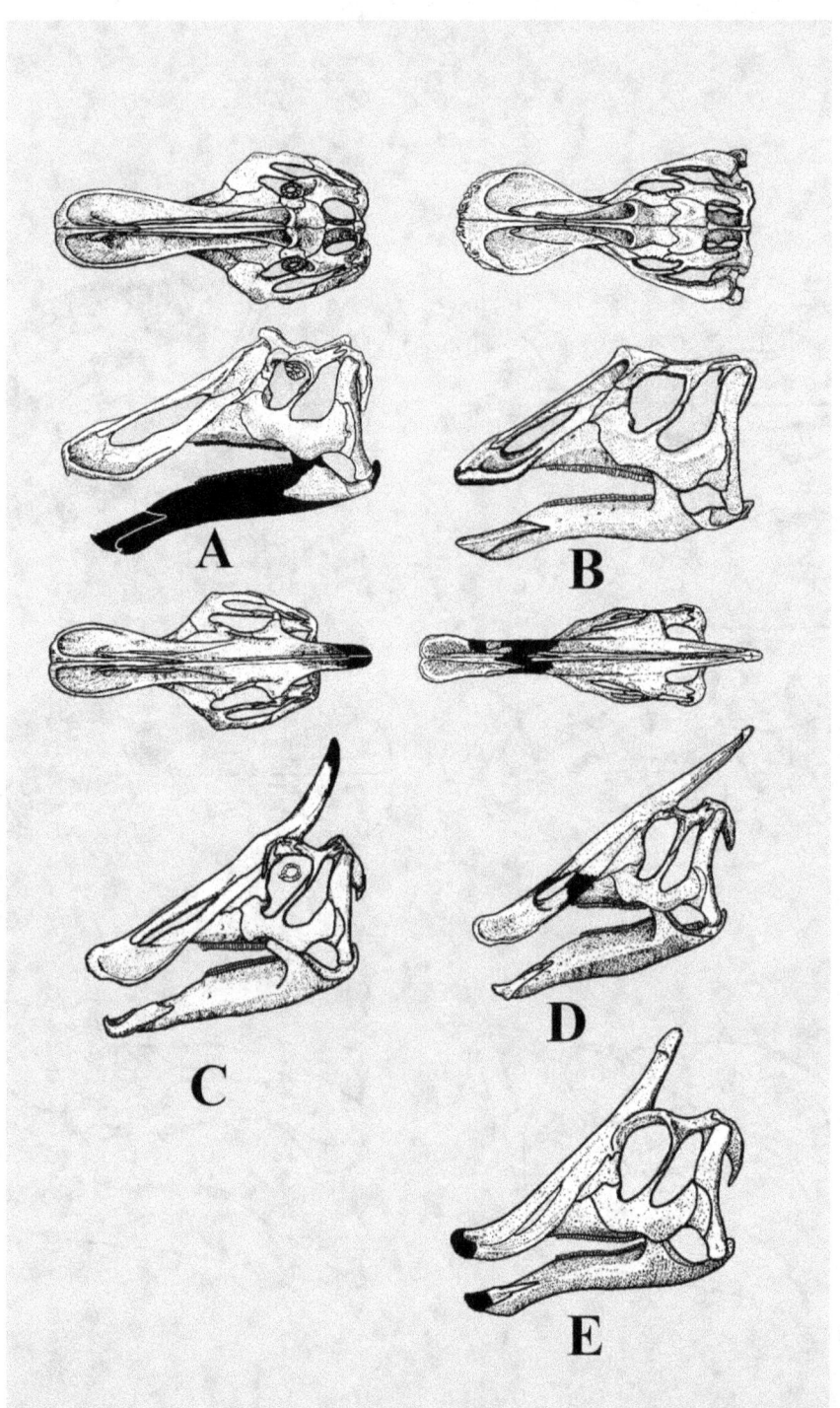

Figure 4). Saurolophinae Hadrosaurids; A) *Prosaurolophus maximus*, top and side view; B) *Prosaurolophus blackfeetensis*, top and side view; C) *Saurolophus osborni*, side and top view; D) *Saurolophus angustirostris*, top and side view of adult, bottom, side view of a juvenile.

Bibliography

Bartholomai, A., and Molnar R. E., 1981, *Muttaburrasaurus*, a new Iguanodontid (Ornithischia: Ornithopoda) Dinosaur from the Lower Cretaceous of Queensland: Memories of the Queensland Museum, v. 20, n. 2, p. 319-349.

Horner, J., R., 1983, Cranial osteology and morphology of the type specimen of *Maiasaura peeblesorum* (Ornithischia: Hadrosauridae), with a discussion of its phylogentic position: Journal of Vertebrate Paleontology, v. 3, n. 1, p. 29-38.

Horner, J. R., 1992, Cranial morphology of *Prosaurolophus* (Ornithischia: Hadrosauridae) with descriptions of two new hadrosaurid species and an evaluation of hadrosaurid phylogenetic relationships: Museum of the Rockies Occasional Paper n. 2, p. 1-119.

Hunt, A. P., and Lucas S. G., 1993, Cretaceous Vertebrates of New Mexico: In: Vertebrate Paleontology in New Mexico, New Mexico Museum of Natural History and Science Bulletin, n. 2, p. 76-91.

Lambe, L. M., 1914, On *Gryposaurus notabilis*, a new genus and species of trachodont dinosaur from the Belly River Formation of Alberta, with a description of the skull of *Chasmosaurus belli*: The Ottawa Naturalist, v. 27, n. 11, p. 145-153.

Langston, W. Jr., 1974, Nonmammalian comanchean tetrapods: Geoscience and Man, v. 8, p. 77-102.

Lull, R. S., and Wright N. E., 1942, Hadrosaurian Dinosaurs of North America: The Geological Society of America, Special Paper, n. 40, p. 1-242.

Molnar, R. E., 1996, Observations of the Australian Ornithopod Dinosaur, *Muttaburrasaurus*: Gondwana Dinosaurs of India: Affinities and Palaeobiography. Proceedings of the Gondwana Dinosaur Symposium, Memoirs of the Queensland Museum, v. 39, part 3, p. 639-652.

Norman, D. B., 1998, On Asian ornithopods (Dinosauria: Ornithischia). 3. A new species of iguanodontid dinosaur: In: A study of fossil vertebrates. Edited by. Norman D. B., Milner A. R., and Milner A. C., Zoological Journal of Linnean Society, v. 122, p. 291-348.

Ostrom, J. H., 1970, Stratigraphy and Paleontology of the Cloverly Formation (Lower Cretaceous) of the Bighorn Basin Area, Wyoming and Montana: Peabody Museum of Natural History, Yale University, Bulletin 35, p. 1-234.

Rozhdestvensky, A. K., 1957, A duckbilled dinosaur, a sauroloph from the Upper Cretaceous of Mongolia: Vertebrata PalAsiatic, v. 1, n. 2, p. 129-149.

Rozhdestvensky, A. K., 1968, Hadrosauridae of Kazakhstan: In: Upper Paleozoic and Mesozoic Amphibians and Reptiles in the USSR, Nauka. Moscow, p. 97-141.

Sternberg, C. M., 1953, A new hadrosaur from the Oldman Formation of Alberta, discussion of nomenclature: Bulletin of the National Museum of Canada, v. 128, p. 275-286.

Taquet, P., 1976, Geologie et Paleontologie du gisement de Gadoufaoua (Aptian du Niger): Cahires Paleont, 191pp.

Trexler, D. L., 1994, A new specimen of *Maiasaura* from the Two Medicine Formation, Montana, and a diagnostic revision of the genus: Journal of Vertebrate Paleontology, v. 14, supplement to n. 3, Abstracts of Papers, Fifty-Fourth Annual Meeting, Society of Vertebrate Paleontology, Burke Museum, University of Washington, Seattle, Washington 7, September 1994, p.50A.

Winkler, D. A., Murry P., and Jacobs L. L., 1997, A new species of *Tenontosaurus* (Dinosauria: Ornithopoda) from the Early Cretaceous of Texas: Journal of Vertebrate Paleontology, v. 17, n. 2, p. 330-343.

For the Dinosaur Collector and Enthusiast

PREHISTORIC TIMES

August/September 1999   No. 37

The PT
Interview:
Jan Sovak

Pyro Model Kits

Elasmosaurus

Ford, T. L., 1999-2000, How to Draw Dinosaurs. Roman noses and Cassowary crests: Ornithopoda heads. Part 2: Lambeosaur hadrosaurs: Prehistoric Times, n. 39, p. 14-15.

Chapter 19

Roman noses and Cassowary crests: Ornithopoda heads. Part 2: Lambeosaur hadrosaurs

Part 2: Lambeosaur hadrosaurs.

Before I start on the lambeosaurids I must bring up something that I learned on my recent summer dino trip (the summer of 2000). I was at the Eccles Dinosaur Park in Ogden, Utah, waiting to get in when I saw an *Edmontosaurus* skull cast. The skull was facing me when I noticed that the so-called 'flat' headed dinosaurs aren't really flat headed, that they do have ornamentation on the skull. The ornamentation is, is the orbits themselves. The orbit in *Edmontosaurus* is very high. In front view/cross section this area forms a "U". The actually position of the eye would be in the middle of the orbit and not at the top. I asked Mike Brett-Surman about this (he is one of the worlds experts on hadrosaurs) and he said that yes, they do have a raised orbit area. He said that area may have been very colorful, or had a small ridge of scutes or horns, etc, making this area more 'attractive'. *Anatotitan, Shantungosaurus* had relatively short eye ridges, with *Edmontosaurus* having the largest. (figure 1).

There are four genera of lambeosaur hadrosaurs that have crests (**Editors note: this article was written 15 years ago and in that time several new crested lambeosaurs have been described**), *Corythosaurus, Lambeosaurus Hypacrosaurus* and *Parasaurolophus***,** with the latter being the most famous and popular. I won't be going into the function of the crest/nasal passages, since it doesn't pertain to the topic of this article. Helmeted lambeosaurs' crest isn't thin in cross-section, as often depicted (figure 2). The very top of the crest is thin, then it expands outward to accommodate the nasal passages inside. *Lambeosaurus'* small 'handle' also isn't thin (figure 4). Seen from above the widest part of the handle is where it joins the crest, while the end of the handle is the thinnest. In *Parasaurolophus* the crest is oval in shape (figure 5).

There have been 7 species named under the genus *Corythosaurus* (helmet crest). All have been referred to *C. casuarius*, some being juveniles, and some females and males. All are from the Dinosaur Park Formation, in Alberta Canada (same as *Gryposaurus*) (see figure 2 for the breakdown of referred males, females, juveniles).

Hypacrosaurus also has a helmet crest, but the crest is smaller than *Corythosaurus* respectively. The first *Hypacrosaurus* named, *H. altispinus*, is one of the latest lambeosaurs while *H. stebingeri* comes from an earlier age. *H. stebingeri* has a longer skull and is known from nests, eggs, juveniles and adults. (see figure 3 for the breakdown of referred males and juveniles).

*Lambeosaurus lambei***,** or, the hatchet headed lambeosaur comes from the same place and horizon as *Corythosaurus*. *Lambeosaurus magnicristatus* (also from the same place as *Corythosaurus*), has the largest crest of any of the lambeosaurs. The crest nearly hangs over to the tip of the snout. (see figure 4 for the breakdown of referred males, females, and juveniles).

Parasaurolophus has the shortest 'front' end of the skull with the longest crest. *P. walkeri* is from the same area as *Corythosaurus* and is the most complete specimen of *Parasaurolophus*. *P. tubicen* also has a long crest and is from New Mexico. It is known from a fragmentary skull. Robert Sullivan has recently described a new specimen of *P. tubicen*. It also comes from New Mexico and has a 'thicker' crest than the type. *P. crytocristatus* has a shorter crest and is also known from a chunk of crest from Utah. It has been assumed that *P. crytocristatus* is a female of *P. tubicen*. Robert Sullivan and others have disputed this, and they come from different ages. (figure 5). (**Editors note: After more research, I no longer believe *P. crytocristatus* is a *Parasaurolophus*. Its premaxillary is longer than other *Parasaurolophus* specimens. This gives the skull a long 'face'. It may in fact be a new genus.)**

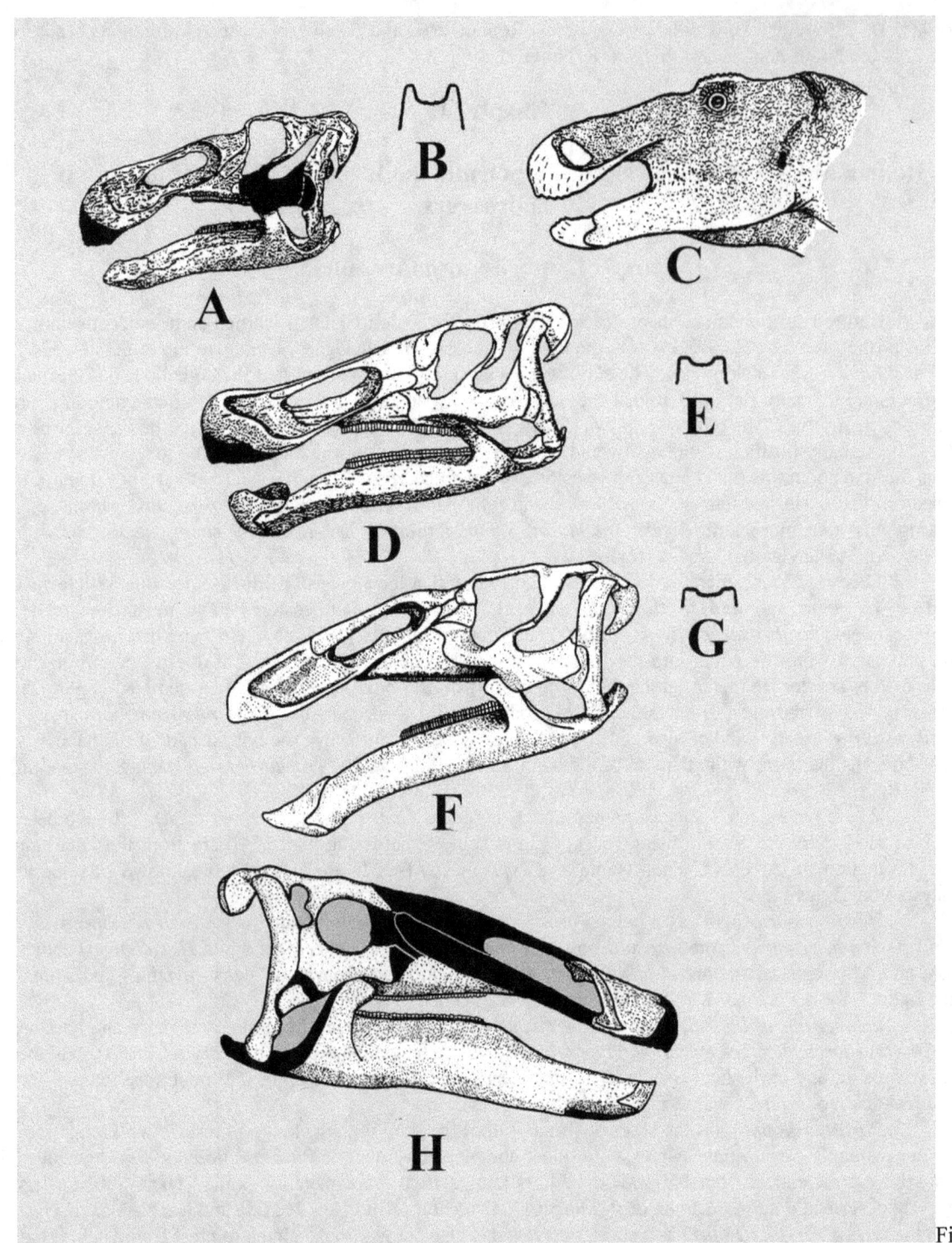

gure 1). Hadrosaurid hadrosaurs showing the 'U' shaped orbits; A) *Edmontosaurus regalis* (NMC 2289) in side view; B). cross-section of the orbits and C) life restoration; D) *Edmontosaurus annectens* (USNM 2424) in side view, E) cross-section of the orbits; F) *Anatotitan copei* (AMNH 5730) in side view (new interpretation) and G) cross-section of the orbits; H) *Shantungosaurus giganteus* in side view.

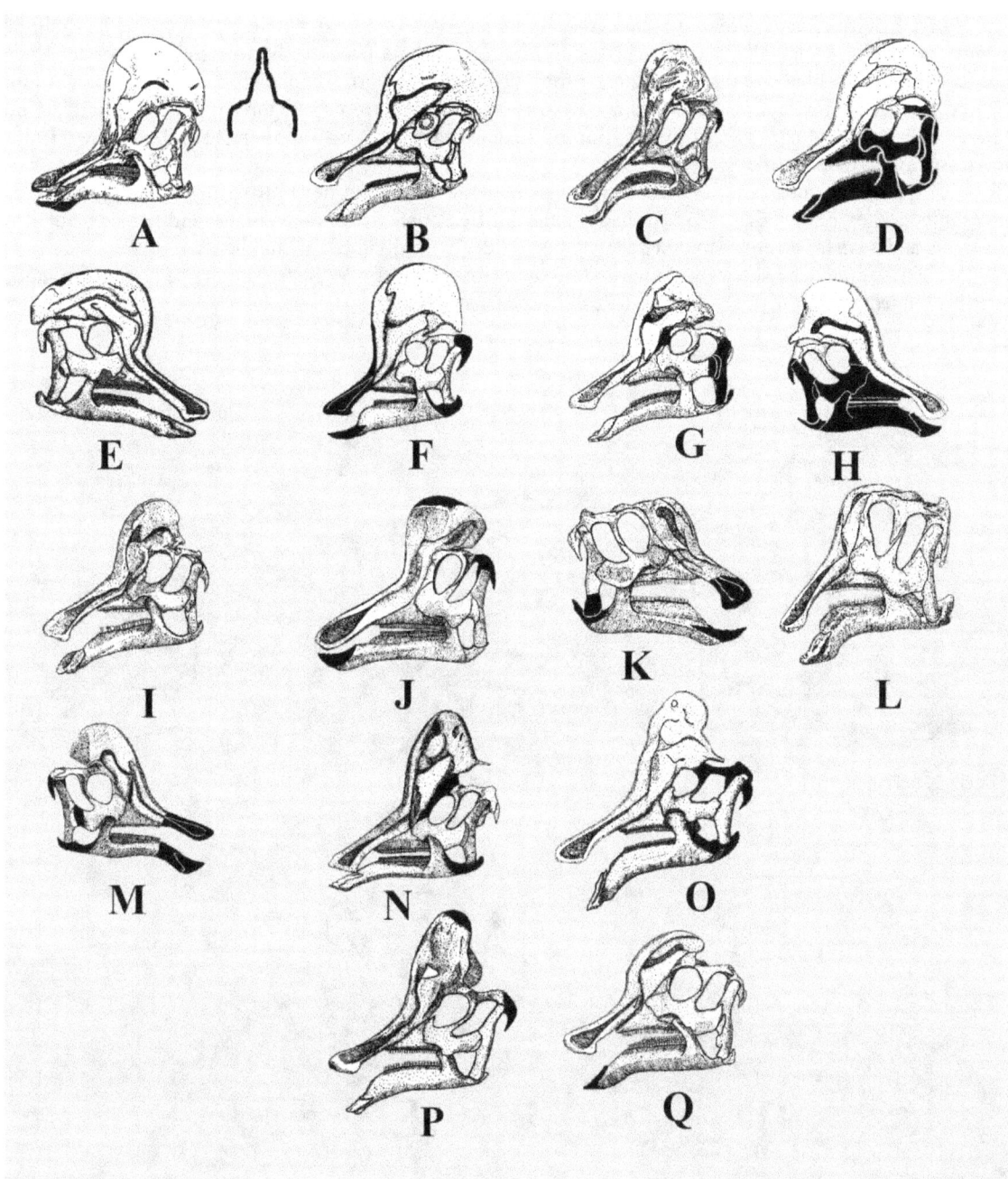

Figure 2). *Corythosaurus casuarius* A-M, *Corythosaurus intermedius* N-Q); Males, A) *C. casuarius*, type AMNH 5240 in side and cross-section, B), AMNH 5338, C) RTMP 80.40.1, D) ROM 1933, E), RTMP 84.121.1, F). ROM 871: Females, G) *C. excavatus*, UA 13, H). NMC 8676, I) ROM 845, J) *C. bicristatus*, ROM 868; K), Juveniles, *Tetragonosaurus (Procheneosaurus) erectofrons*, ROM 759 L), AMNH 5461, M); *T. (P.) cranibrevis*, NMC 8633.; N) *Corythosaurus intermedius*, ROM 776, O) ROM 777, P), ROM 777, and Q) NMC 8704.

What about *Tsintosaurus*? The crest of *Tsintosaurus* is a very controversial subject. The bottom line is this, Eric Buffetaut believes the thin crest is real and did extend like a unicorn's horn, while Philippe Taquet believes it is actually the nasal bones shifted forward and needs to be lowered like normal 'flat' headed hadrosaurs. I believe that Taquet is correct. The tip of the horn, flipped around, looks like it would fit into the premaxilla and be very similar to the hadrosaurid hadrosaurs (figure 6). (**Editors note: I no longer believe this, due to recent publication by Prieto-Marquez, and Wagner, 2013, had shown it did have a near vertical crest**).

Why all the crest shapes? Dodson, Weishampel, and Hopsen have all written several papers, on this subject. They believe that the crest shapes helped differentiate not only the males and females, but also juveniles and adults. The crest structure also helped to identify individuals within herds.

The varied head shapes were needed because, for the most part, hadrosaurs were herding animals, and this helped identify mates, males, females, juveniles, and genera.

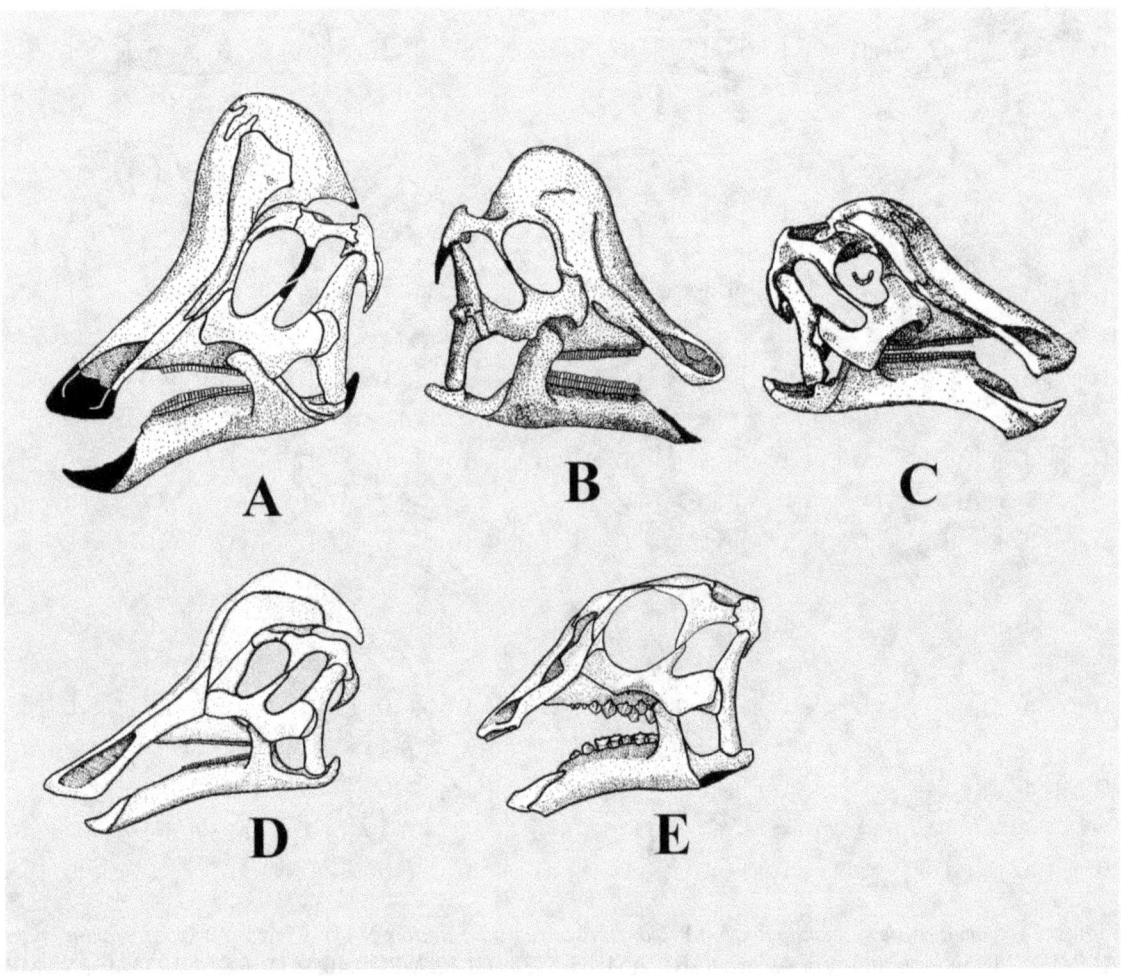

Figure 3). *Hypacrosaurus altispinus*; A) Adult referred skull, NMC 8501, B) Juveniles, RTMP P82.10.8; C) *Cheneosaurus tolmanensis* NMC 2246; D) *H. stebingeri* adult MOR 549, E) hatchling (composite).

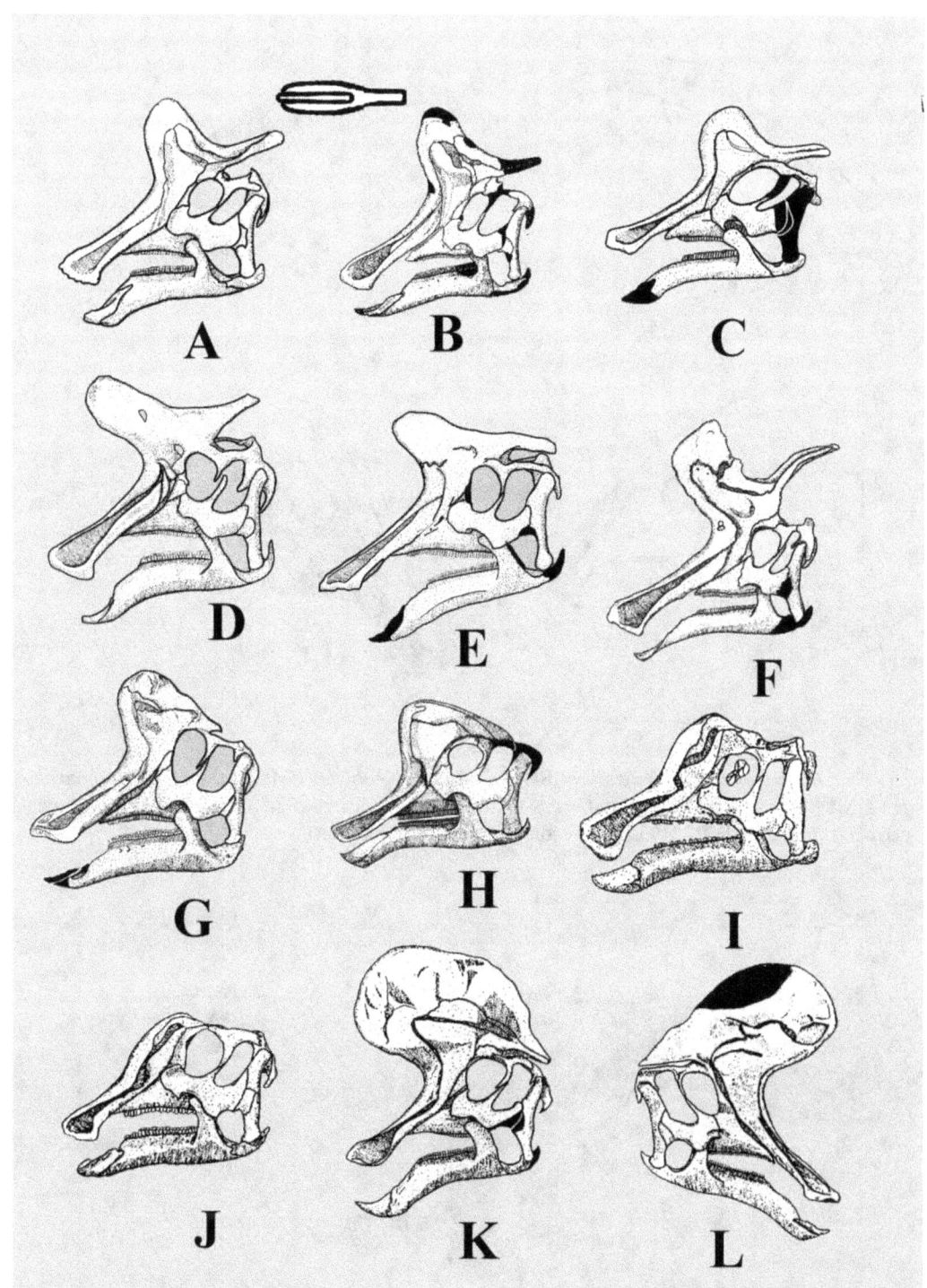

Figure 4). *Lambeosaurus*. Males A) *L. lambei* NMC 2869 in side and dorsal view, B) (*Stephanosauurus marginatus*) NMC 351, C). YPM 3222, D), WMUC 1479, E) RTMP 81.37.1, F) ROM 794; Female *L. clavinitialis* G). NMC 8703; Juvenile female H) *Corythosaurus frontalis* ROM 874; Juveniles I) *Tetragonosaurus praeceps* AMNH 5340, J) ROM 758: *L. magnicristatus* K) NMC 8705, L) PMAA P.66.4.

137

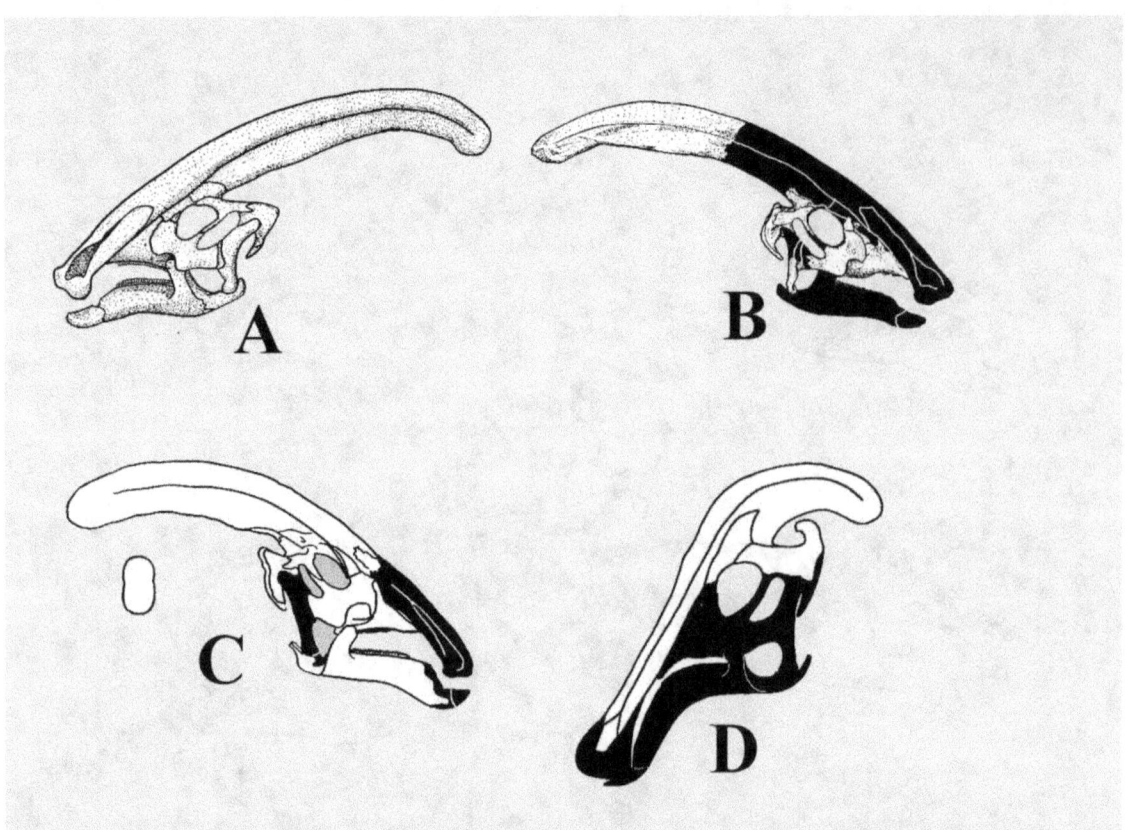

Figure 5). *Parasaurolophus*. A) *P. walkeri* ROM 768, B) *P. tubicen* PMU R 222 C) *P. tubicen* referred specimen NMMNH P-25100 in side and cross-section,. (all modified after Sullivan & Williamson, 1999); D) *P. cyrtocristatus* FMNH P27393 (new reconstruction/interpretation)

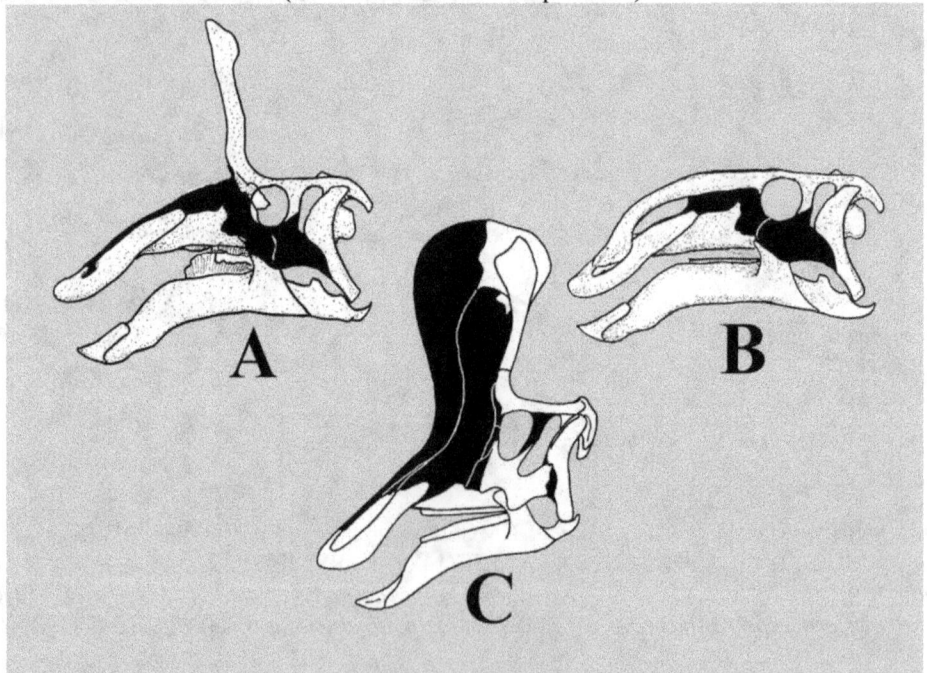

Figure 6). *Tsintaosaurus spinorhinus* A) with the 'unicorn' horn, B) like a flat headed hadrosaur; and C) new interpretation modified from Prieto-Marquez, and Wagner, 2013).

Stratigraphic ages and localities.

Corythosaurus is from Dinosaur Provincial Park, Alberta Canada. Dinosaur Park Formation, Judith River Group, Campanian, Late Cretaceous. C. casuarius is from the lower section of the formation and C. intermedius is from the upper section of the formation.

Hypacrosaurus altispinus is from near Drumheller, Alberta, Canada. Edmonton Group, Horseshoe Canyon Formation, Early Maestrichtian, Late Cretaceous. *H. stebingeri* is from Badger Creek, Glacier County, Montana, Judith River Group, Upper Two Medicine Formation, Campanian, Late Cretaceous and Devil's Coulee, Southern Alberta, Canada, Judith River Group, Campanian, Late Cretaceous.

Lambeosaurus lambei & L. magnicristatus is from Dinosaur Provincial Park, Alberta Canada. Dinosaur Park Formation, Judith River Group, Campanian, Late Cretaceous.

Parasaurolophus walkeri is from Dinosaur Provincial Park, Alberta Canada. Dinosaur Park Formation, Judith River Group, Campanian, Late Cretaceous. *P. tubicen* is from Barrel Spring Arroyo and Hunter Wash, San Juan County, New Mexico and is from the De-na-zin Member of the Kirtland Formation, Late Maestrichtian, Late Cretaceous. *P. cyrtocristatus* is from near Coal Creek, McKinley County, New Mexico, Fruitland Formation, Campanian, Late Cretaceous and from Henryville Creek, Garfield County, New Mexico, Kaiparowits Formation, Maestrichtian, Late Cretaceous.

Tsintaosaurus spinorhinus is from Laiyang Basin, Shantung, China, Wangshih Series, Campanian, Late Cretaceous.

Biblography;

Buffetaut, E., and Tong-Buffetaut H., 1993, *Tsintaosaurus spinorhinus* YOUNG and *Tanius sinensis* Wiman: a preliminary comparative study of two hadrosaurs (Dinosauria) from the Upper Cretaceous of China: Compte rendu hebdomadaire des seances de l'Academie des Sciences Paris, tomo 317, serie 2, p. 1255-1261.

Dodson, P., 1975, Taxonomic implications of relative growth in lambeosaurinae hadrosaurs: Systmatic Zoology, v. 24, p. 37-54.

Hopsen, J. A., 1975, The evolution of cranial display structures in hadrosaurian dinosaurs: Paleobiology, v. 1 n. 1, p. 21-43.

Horner, J. R., and Currie P. J., 1994, Embryonic and neonatal morphology and ontogeny of a new species of *Hypacrosaurus* (Ornithischia, Lambeiosauridae) from Montana and Alberta: In: Dinosaur Eggs and Babies, edited by Carpenter K., Hirsch K. F., and Horner J. R., Cambridge University Press, p. 312-336.

Hu, C. C., 1973, A new Hadrosaur from the Cretaceous of Chucheng, Shantung: Acta Geologica Sinica, n. 2, p. 179-206.

Lull, R. S., and Wright N. E., 1942, Hadrosaurian Dinosaurs of North America: The Geological Society of America, Special Paper, n. 40, p. 1-242.

Ostrom, J. H., 1961, A new species of hadrosaurian dinosaur from the Cretaceous of New Mexico: Journal of Palaeontology, v. 35, n. 3, p. 575-577.

Ostrom, J. H., 1963, *Parasaurolophus cyrtocristatus*, a Crested Hadrosaurian Dinosaur from New Mexico: Fieldiana, Geology, v. 14, n. 8, p. 143-168.

Prieto-Marquez, A., and Wagner, J. R., 2013, The 'Unicorn' dinosaur that wasn't: a new reconstruction of the crest of *Tsintaosaurus* and the early evolution of the Lambeosaurine crest and rostrum. Public Library of Science (PLOS), One, v. 8, n. 11, 20 pp.

Sullivan, R. M., and Williamson T. E., 1999, A New Skull of *Parasaurolophus* (Dinosauria: Hadrosauridae) from the Kirtland Formation of New Mexico and a Revision of the Genus: New Mexico Museum of Natural History and Science, Bulletin n. 15, p. 1-52.

Taquet, P., 1991, The status of *Tsintaosaurus spinorhinus* YOUNG, 1958 (Dinosauria): In: Fifth Symposium on Mesozoic Terrestrial Ecosystems and Biota, Extended Abstracts, edited by Kielan-Jaworowska Z., Heintz N., and Nakrem H. A., Contributions from the Paleontological Museum, University of Oslo, n. 364, 1991, p. 63-64.

Weishampel, D. B., 1981, The nasal cavity of lambeosaurinae hadrosaurids (Reptilia: Ornithischia): comparative anatomy and homologies: Journal of Paleontology, v. 55, n. 5, p. 1046-1057.

Young, C.-C., 1958, The Dinosaurian Remains of Laiyang, Shantung: Palaeontologia Sincia, Whole Number 142, new series C, n. 16, p. 1-138.

For the Dinosaur Collector and Enthusiast

PREHISTORIC TIMES

Feb-Mar 2000   No. 40

Paleonews 1999
The top dinosaur discoveries of the year

The PT Interviews:
Martin Lockley
Michael Triebold

and much more!

40th BIG ISSUE!

© R DELGADO
10-99

U.S. $5.95 • Canada $6.95
02>

0 74470 92325 1

140

Ford, T. L., 2000, How to Draw Dinosaurs. The theropod nose: Prehistoric Times, n. 40, p. 14.

Chapter 20

The theropod nose

The easiest way for an artist to depict the nose in theropods is to make the snout round like lizards. Looking at the skull of a theropod in side view it's easy to understand why. The nasals often look flush with the skull either from bad lighting in photos, or because an illustration isn't detailed enough. Actually the nasals are inset from the snout, and that the dorsal, nasal area itself looks like a rod or handle. The only known exceptions are the abelisaurids *Carnotaurus* and *Majungasaurus* where this area is thicker. The rod like nose is also true for most ornithopods and sauropods, but I'll be talking about those in a later paper.

The best way to explain is for me to section the nasal area of *Allosaurus*. I'll be doing this via illustration (since I can't really cut up a skull). The nasal/snout area is comprised of a paired premaxillae, nasals and to a small degree the dorsal area of the maxillae. The premaxilla connects to the nasal via thin prong-like projections at both the posterior dorsal and ventral side. These projections are thin and are easily broken off during fossilization or collection. The front end of the premaxilla is rounded to various degrees in theropods. In *Dilophosaurus* and spinosaurids the premaxilla is tear dropped shaped in dorsal view and tapers to the posterior end. The premaxilla and nasals contact each other which precludes the maxilla so that it does not contact the nose itself. In side view the skull of *Tyrannosaurus rex's* maxilla seems to contact the nose. The reason for this is that the snout is the widest in any theropod and the maxilla has grown taller while the premaxilla and nasal have become more like shelves and flattened at the connection of the lower edge of the premaxilla, nasal and maxilla area.

The tip of the premaxilla is rounded, and at about half the height, the rod-like projection of the premaxilla begins. The ventral projection curves into the center of the naries and starts a sort of shelf in the area of the nose. This area can be either flat or slanted. The upper part of the premaxilla that connects to the nasals has a rod-like construction that continues backward to about two thirds of the nose, then begins to expand. The widest part of the nose is at the posterior end. Not all theropods have the naries at the tip of the premaxilla. In dilophosaurids, spinosaurids, compsognathids, and dromaeosaurids to name a few, the naries is inset from the tip of the naries.

As the animal breathed the air flowed into the skull at the posterior end of the nose, through the nose itself as well as into pneumatic 'tubes'. This can be best seen in the *Tyrannosaurus* 'Stan', which has the pneumatic areas of the maxilla prepared out of the matrix. The ventral side of the naries is open to the palate, which was most probably covered with soft tissue. Was there a piece of cartilage, or septum, splitting the naries in half? I don't think there was enough room for one, or ventral side of the naries wouldn't have been able to contain the septum. I may be wrong, but I think that the septum needed some bony support and the naries of theropods lack this.

I've asked several paleontologist's about whether or not the nose would have been an 'open' hole, similar to sea gulls and some vultures, or more like a lizard's. They didn't really know but thought they may have had a septum. I'm more for the 'like a gulls' theory. As stated before, I don't think there was enough room on the naries for muscle attachment to cover them.

Some paleontologists have speculated that the nose and top of the skull had a bony covering or sheath like a beak of a bird. If true, the nose could have had a fleshy covering like some predatory birds. This fleshy area is called a 'cere'. In the majority of birds the nasal opening is quite large, but the horny sheath that covers most of it leaving a much smaller opening. Could theropods have had a cere? Or a horny sheath over the naries?

So for now, until research proves otherwise I'll illustrate the nose as an open area. (**Editors note: I no longer believe this and in a latter issue corrected this**).

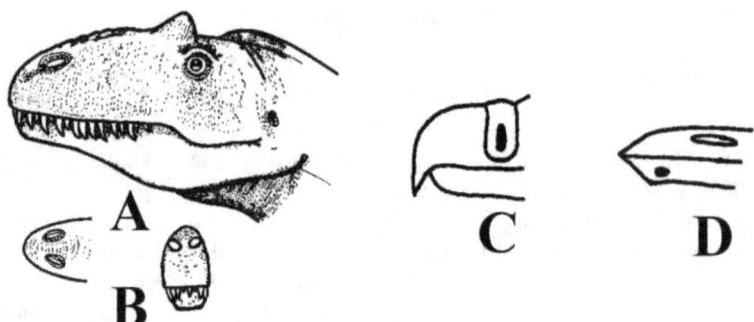

Figure 1). A) Drawing of *Allosaurus* with a 'lizard-like' nose; B) snout of *Allosaurus* in dorsal and lateral view; C) hawk in lateral view; D) sea gull in lateral view.

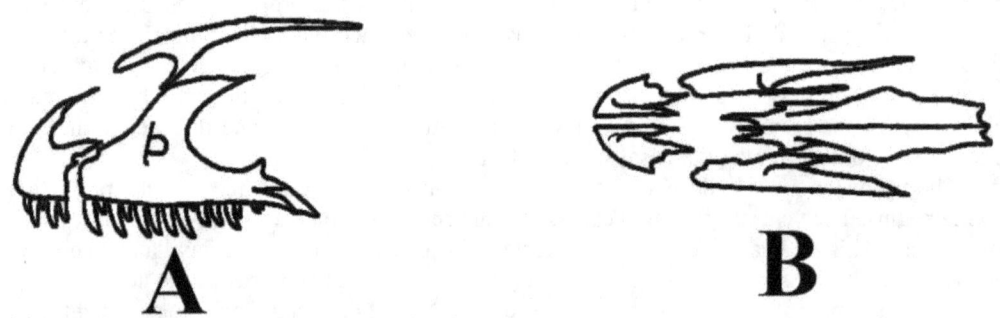

Figure 2). A) Premaxilla; B) nasals; C) maxilla in dorsal and lateral view.

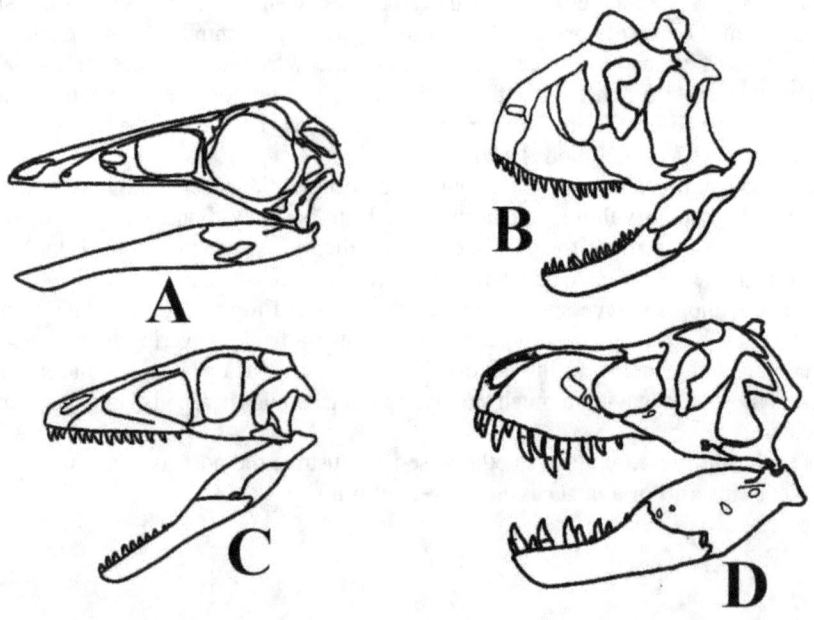

Figure 3). Skulls of theropods; A) *Dromiceiomimus*; B) *Carnotaurus*; C) *Deinonychus*; D) *Tyrannosaurus*.

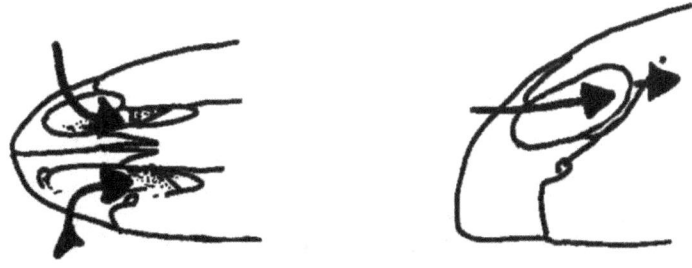

Figure 4). Illustration showing how the air passes into the skull.

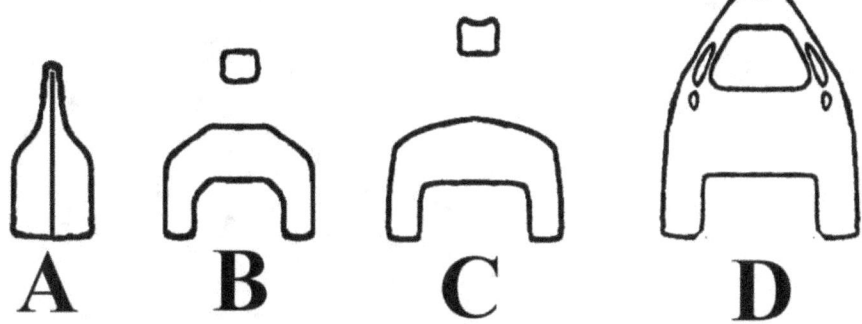

Figure 5). Cross sections of the skull of *Allosaurus*; A) front of snout; B) 2/4ths of the snout; C) 3/4ths of the snout; D) posterior end of snout showing the large opening for the nose and the maxillae pneumatic areas (speculation).

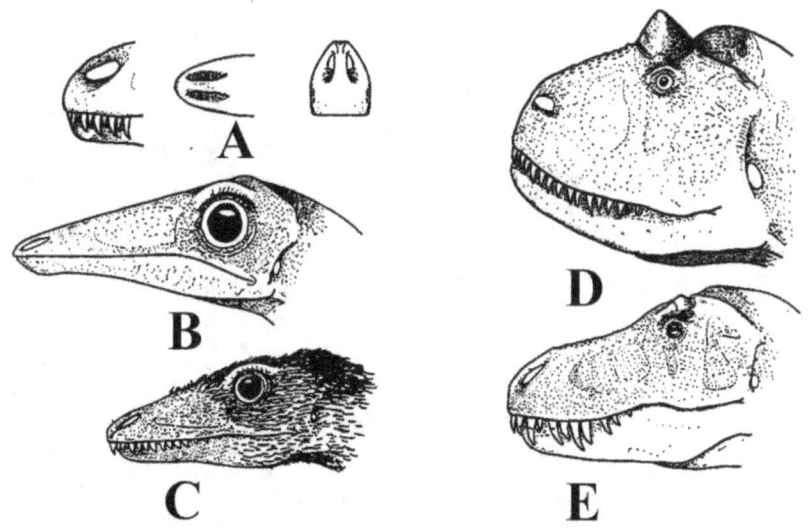

Figure 6). Heads of theropods. A) *Allosaurus* snout in lateral, dorsal and frontal views; B) *Dromiceiomimus*; C) *Carnotaurus*; D) *Deinonychus*; E) *Tyrannosaurus*.

For the Dinosaur Collector and Enthusiast

PREHISTORIC TIMES

April/May 2000   No. 41

Dino Modeling Issue
How-To's, Reviews
Collectibles!
Science, Art
& Much More!

U.S. $5.95 • Canada $6.95
04 >

Disneys' "Dinosaur"

Ford, T. L., 2000, How to Draw Dinosaurs. Scales, spines, sails & scutes, the backs of sauropods: Prehistoric Times, n. 41, p. 14-15.

Chapter 21

Scales, spines, sails & scutes, the backs of sauropods

Sauropods are often thought to have had an elephantine hide - loose fitting skin with no ornamentation on the back or elsewhere. Recently discoveries show that some sauropods actually did have spines or scutes. Science now knows that the backs of sauropods show a wide variety of ornamentation; in fact, more so than other dinosaurs.

What did the skin of sauropods look like? Several specimens with skin impressions intact have been found. Unfortunately, to get to the more valuable skeleton, some of these specimens had the skin literally blasted away. What little skin is known had small oval scales that wouldn't have been readily seen from 10 feet away from the animal. In Argentina an extensive nesting sight of sauropods has been found. Eggs were found; also embryos in eggs, and one neonate included skin impressions. Unfortunately the patch of skin had fallen off then and then sunk to the bottom of the egg, so it is unknown from which part of the animal the skin came from (figure 1).

Stephen Czerkas theorized that diplodocids didn't have whip-like tails. A new specimen that he studied from Howe Quarry (Wyoming) had skin impressions from the tail that included tall, triangular spikes in a single row running down the middle, looking much like a tuatara. These spikes were varied in height and shape. The large area of 'skin' from the scutes to the caudal vertebrae is extensive. If this were the natural state of the animal it wouldn't have had a 'whip-like' tail (figure 1, 2).

Some sauropods had bifurcated neural spines. Would this area been covered in muscle and skin; or just skin? Muscle and sinew would have covered most of the area connecting vertebrae to vertebrae, but I don't think enough to cover the middle of the bifurcated area. Most likely there would have been a 'trough' down the middle of the back. In camarasaurids the cervicals had the bifurcated spines that ended at the first dorsal (figure 2).

Diplodocids and dicraeosaurids had bifurcated spines down to the pelvic vertebrae. I wonder if water could have collected in small pools on the back? *Dicreaosaurus sattleri* had the tallest spines and would have had a short "sail" (figure 2).

Some sauropods had sails. *Rebbachisaurus* and *Rayososaurus tessonei* (a.k.a. *Rebbachisaurus tessonei* from Argentina) had the tallest single spined vertebrae. These spines would have given the sauropod a short sail (figure 3).

Amargasaurus is often thought to have had a bifurcated sail. As I mentioned in PT (volume 27) I and others (others being Argentine paleontologist), believe that the tall thin, rod-like vertebral spines didn't end in a sail but has long thin spikes. It is important to note that the axis has a single long spine with the following vertebrae bifurcated.

Titanosaurids are the only sauropods discovered so far that had scutes. Not all titanosaurids had scutes preserved, however. No titanosaur scutes are known from North America (**Editors note: a recent a paper [2015] described *Alamosaurus* dermal scutes from North America**). How the scutes in the armored titanosaurids were arranged is difficult to know and is open to speculation. For the most part they are thick oval objects with a short ridge down the middle and highly vascular.

Other scutes had a short rounded point. Mark Hallet, the first to paint an armored sauropod for George Olshevsky's Science Digest article littered the back with scutes. Some Argentine paleontologist believe that titanosaurs had a double row down the middle of the back; some think it was a single row. Until an articulated specimen is found, a definite scute pattern is open to discussion (figure 4).

Which brings us to the last sauropod, *Agustinia*. I first drew *Agustinia* in PT from a verbal description. Recently the formal article has been published. This sauropod had a strange array of osteoderms or scutes. The first cervical scutes are spade-like with the broadest surface facing the head and tail. They then take on another shape entirely at about the middle of the neck: The cervical spines appear more like spikes arranged almost horizontally out from the spine with long thin oval spikes. Some think these scutes are either ankylosaurian or stegosaurian but they don't share any similarities to either of the armored dinosaurs and I am in agreement that these belong to this amazing sauropod, *Agustinia* (figure 5).

(Editors note: **I have learned that these spines are now believed to be ribs, and *Agustina* was not armored in any way**).

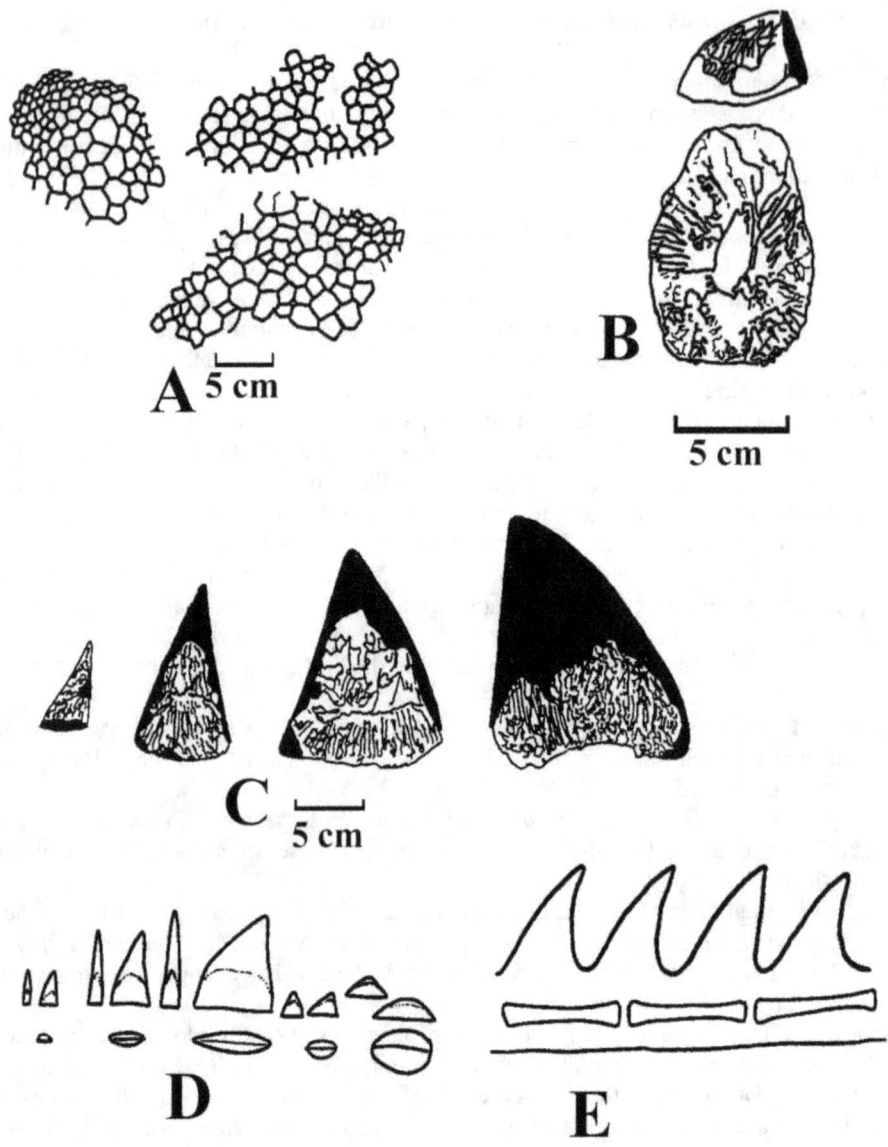

1) Skin impressions of sauropods: A) British Museum No. R. 1870 from the front legs; B) Dinosaur National Monument specimen from the side of neck; C) How Quarry specimen from the belly region; D) different spikes from the How Quarry Specimen in side, top and front view; E) side view of tail with spikes in place. All after Czerkas, 1994.

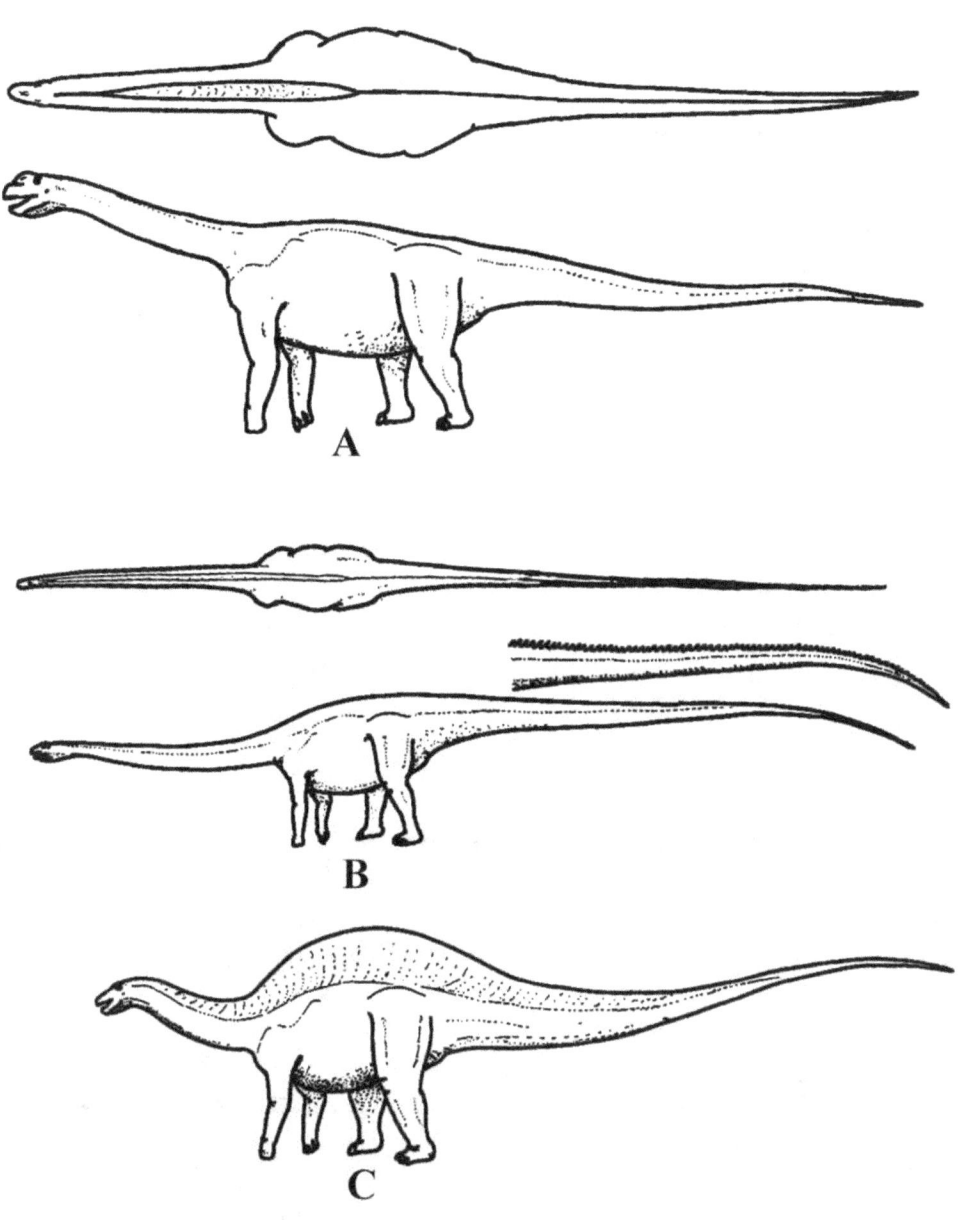

2). A) Top and side view of a *Camarasaurus* showing the bifurcated neural spines; B) Top and side view of *Diplodocus*, and tail section showing the spikes; C) Side view of *Dicraeosaurus sattleri*.

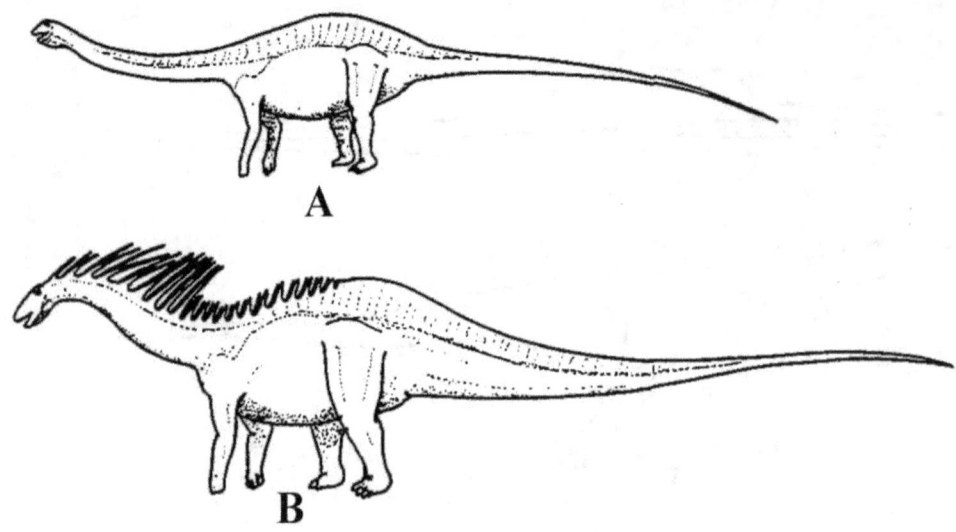

3). A) *Rayosaurus tessonei* in side view; and B) *Amaragasaurus* in side view.

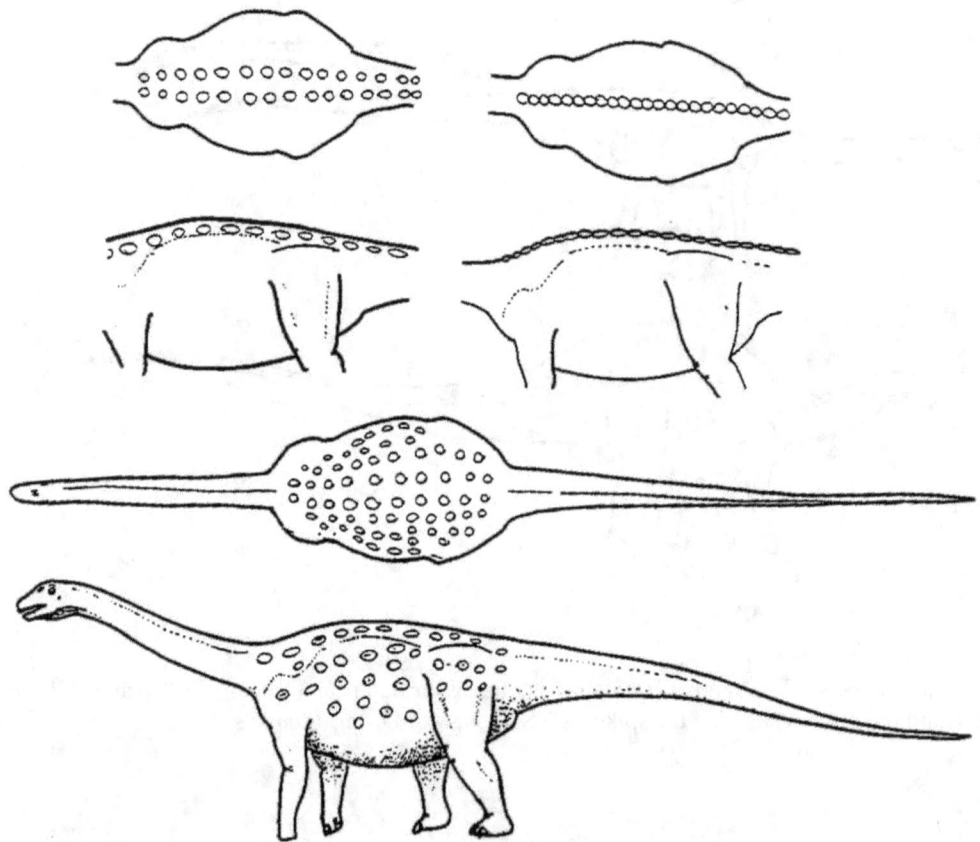

4). Titanosaurid showing the different patterns of scutes; back littered, a double row and a single row.

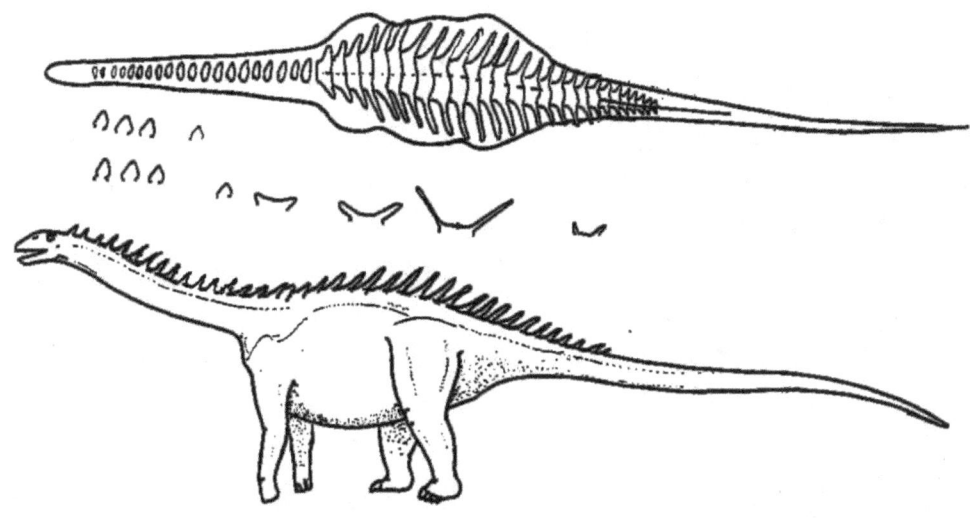

5). *Agustinia* in top and side view.

Bibliography:

Bonaparte, J. F., 1998, Los Dinosaurios de la Patagonia Argentina: Museo Argentino de Ciencias Naturales, p. 46pp.

Bonaparte, J. F., 1999, An Armoured Sauropod from the Aptian of Northern Patagonia, Argentina: In: Proceedings of the Second Gondwanan Dinosaur Symposium, edited by Tomida Y., Rich T. H., and Vickers-Rich P., p. 1-12.

Czerkas, S. A., 1994, The history and interpretation of Sauropod skin impressions: In: Aspects of Sauropod Paleobiology, edited by Lockley M. G., Santos V. F. dos, Meyer C. A., and Hunt A., Revista de Geociencias, Gaia, n. 10, p. 173-182.

Olshevsky, G., 1981, Dinosaur Renaissance: Science Digest, v. 89, n. 7, p. 34-43.

For the Dinosaur Collector and Enthusiast

PREHISTORIC TIMES

June/July 2000   No. 42

In This Issue:

History of
Dino Art

Dino Movies

The Latest
Dino Model Kits

and
much
more!

U.S. $5.95 • Canada $6.95

Ford, T. L., 2000, How to Draw Dinosaurs. Don't wake the sleeping dino: Prehistoric Times, n. 42, p. 14.

Chapter 22

Don't wake the sleeping Dino's

How did dinosaurs sleep? This is a fundamental and intriguing question; one that I have thought about and often been asked about. Did they lie down like big cats (as has been depicted in art), or crocodiles or birds?

The first fantastic painting that Mike Skreptnick showed me (and I will always remember it) was of a *Compsognathus* lying under a tree like a sprawled out cat (There is a sculpture of an *Allosaurus* in the same pose). Was this physically possible? Sure some poses look cool, but what does the skeleton suggest?

The skeleton of dinosaurs are far different from mammals. Mammals have two different types of dorsal vertebrae: the thoracic (the back) where the ribs attach, and a lumbar region without ribs (between the ribs and pelvis) (figure 1A). The backs of mammals are more mobile which allows them to bend and twist. When at rest, cats and dogs (for example) can twist their backs from vertical (front end) to horizontal (hind end). Dinosaurs on the other hand have only a thoracic region (figure 1B). The vertebrae interlock, and in some cases, have tendons (which have ossified during fossilization) to strengthen their back. It has been a long held believe that the tendons would stiffen the vertebrae to a point of near immobility, like the tail in dromaeosaurids.

Recent finds show that there was mobility in the tail, even with the tendons (but not as much as seen in Jurassic Park's Lost World); and this would be true for the back as well. The backs of dinosaurs couldn't twist like mammals. Also we need to remember the pubes. The pubes is long and would extend the lower part of the animal close to where the knee would be if it laid on its side and would hinder the animal from getting up. This is true for all bipedal dinosaurs, whether theropods or ornithischians.

Then how did they sleep or rest? Is there any fossil evidence to help show this? There is both ichnological (footprint) and skeletal evidence that help indicate how they rested. Tracks of a bipedal dinosaur with pubes and ischium marks are known from the Lower Jurassic, Lily Pond Quarry at Gill, Massachusetts (Amherst collection 1/7). It is now thought to have been made by a small theropod. The feet and metatarsal (heels) are flat on the ground, the pubes is near one of the feet and the ischium mark is behind the feet (figure 2A). Gierlinski believes that there are also feather impressions, but that is a topic for another article. Other tracks also show dinosaurs sitting in a similar manner (figure 2B). This indicates that bipedal dinosaurs sat like birds not like cats.

Skeletal evidence can be found in complete or nearly complete articulated skeletons. Recent finds of brooding oviraptors portray the feet under the body and the arms out stretched over their eggs (figure 3A). *Sinornithoides youngi*, a small troodontid from Mongolia, was found in a similar position with the tail wrapped around its body (figure 3B).

I started thinking about resting dinosaurs recently because of a different dinosaur found in the same position as the "brooding" *Oviraptor*. The type specimen of *Psittacosaurus mongoliensis* is in what could be called a "brooding" pose (figure 3C). The hind legs and feet are under the body, the hands outstretched with the palm facing up, and the tip of the tail is in the left palm. It wasn't found lying on eggs or a nest. Since its not a theropod it isn't thought to be brooding (which doesn't mean it couldn't). This leaves two questions; one, if it isn't brooding, what was it doing". Secondly and more importantly, is the specimen in a natural behavioral pose? I believe the answer to the latter is yes and the former is that it is resting. This resting pose has also been found in quadrapedal dinosaurs.

An articulated *Protoceratops* specimen at the American Museum of Natural History, was found in a similar pose. It is lying with its hind legs under its body and its claws perpendicular to the foot (as if digging in). The hands are resting next to the body with the palms up (figure 3D). Crocodiles will rest in a similar manner. Another *Protoceratops* specimen is in a 'curled up' position.

Small quadrapedal dinosaurs probably rested like crocodilians, which sometimes rest arms next to the body with the palms up or down, and with the arms outstretched. Larger quadrapedal dinosaurs, stegosaurs, ceratopians, ankylosaurs, etc., probably laid on their stomachs and NEVER on their sides. Sauropods were too large and probably slept standing up like an elephant. Possibly with one eye open in case of a predators attack. Figure 4 shows how some dinosaurs may have rested.

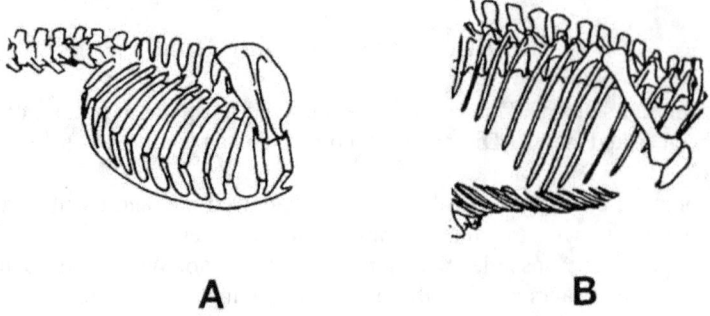

Figure 1). A) Body region of a cat showing the thoracic (rib) section and the lumbar (ribless section) and pelvis; B) *Allosaurus* body region showing the lack of lumbar region and pelvis.

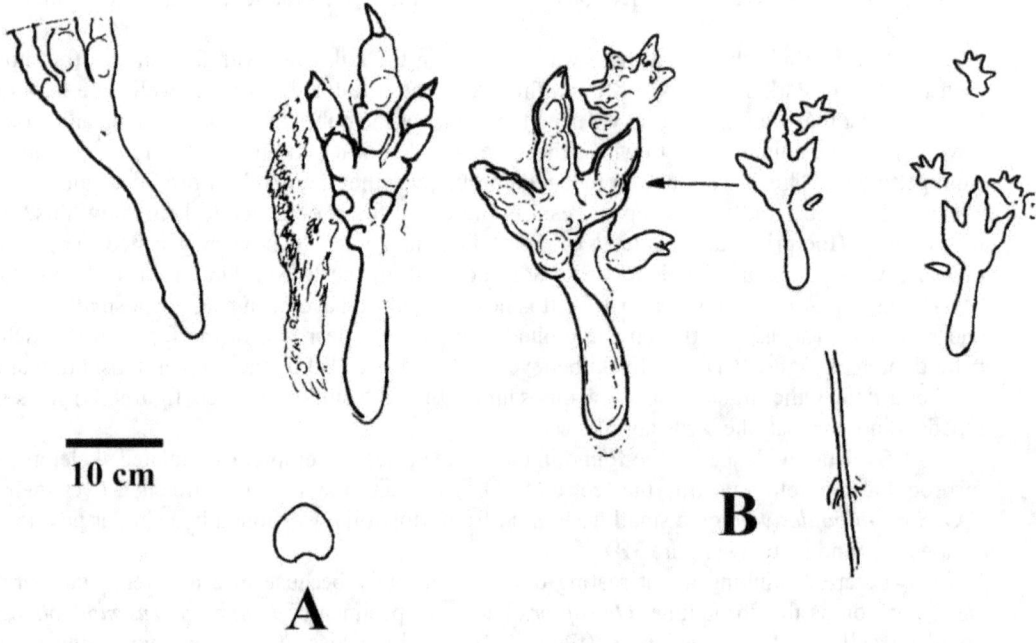

10 cm

Figure 2). A) Amherst collection 1/7; B) *Moyenisauropus natator*; C) side and dorsal view of 1/7.

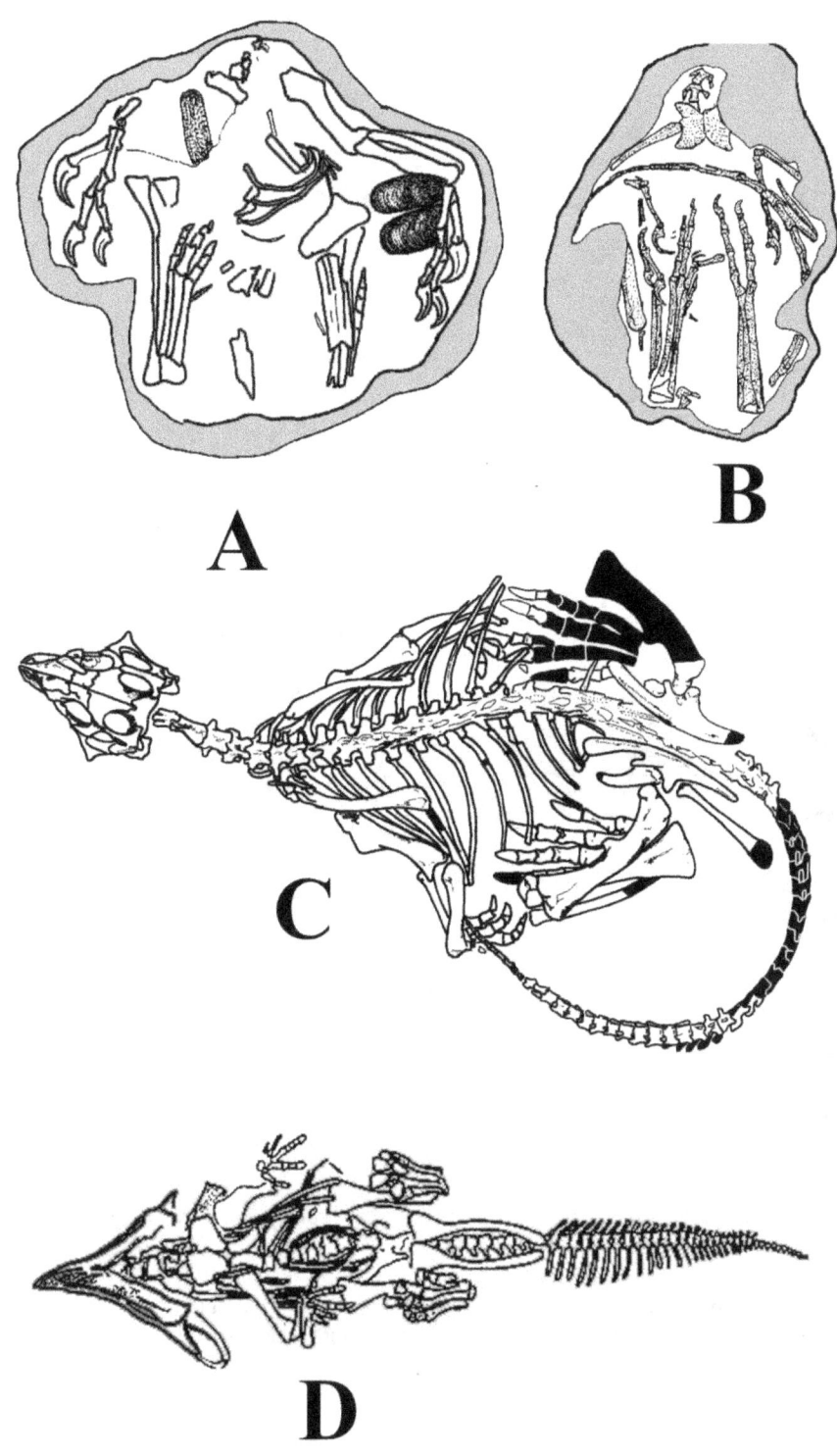

Figure 3). A). Brooding *Oviraptor*; B) *Sinornithoides youngi*; C) *Psittacosaurus* and D) *Protoceratops*.

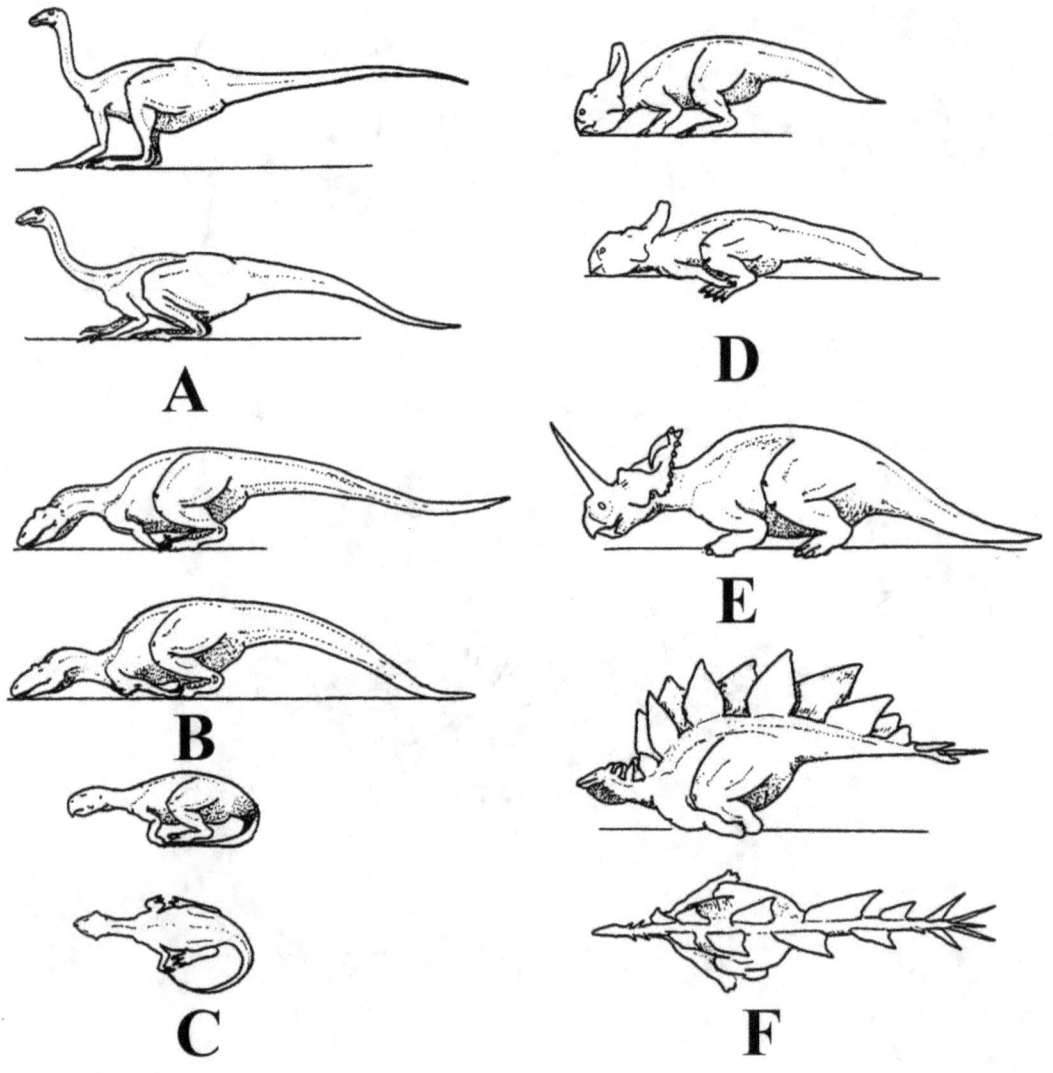

Figure 4). A) *Struithomimus*; B) *Allosaurus*; C) *Psittacosaurus*; D) *Protoceratops*; E) *Centrosaurus* and F) *Stegosaurus*.

To answer the question, bipedal dinosaurs couldn't lie down like a cat, and hunkered down like a bird. Quadrapedal dinosaurs probably rested like crocodilians and sauropods probably slept sanding up like an elephant.

Bibliography

Brown, B. B., and Schlaikjer E. M., 1940, The structure and relationships of *Protoceratops*: Annual of the New York Academy of Science, v. 11, article 3, p. 133-266.

Gierlinski, G., 1996, Feather-like Impressions in a Theropod Resting Trace from the Lower Jurassic of Massachusetts: In: The Continental Jurassic, edited by Morales M., 1996, Museum of Northern Arizona Bulletin, n. 60, p. 179-184.

Norrell, M. A., Clark J. M., Chiappe L. M., and Dashzeveg D., 1995, A nesting dinosaur: Nature, v. 378, p. 774-776.

Osborn, H. F., 1924, *Psittacosaurus* and *Protiguandon*: Two Lower Cretaceous Iguanodonts from Mongolia: American Museum Novitiates, n. 127, p. 1-15.

Russell, D. A., and Dong Z.-M., 1993, A nearly complete skeleton of a new troodontid dinosaur from the Early Cretaceous of the Ordos Basin, Inner Mongolia, People's Republic of China: In: Results from the Sino-Canadian Dinosaur Project. Canadian Journal of Earth Sciences, v. 30, p. 2163-2173.

For the Dinosaur Collector and Enthusiast

PREHISTORIC
TIMES

Aug-Sept 2000 · No. 43

Stephen & Sylvia
Czerkas

The T. Rex
"SUE"

Articles, Interviews
& Reviews
From the World
of Dinosaurs

U.S. $5.95 • Canada $6.95

156

Ford, T. L., 2000, How to Draw Dinosaurs. Theropod feet: Prehistoric Times, n. 43, p. 14.

Chapter 23

Theropod Feet

The topic of this issue was suggested to me by PT reader Fred Crowe. He sent it to me via e-mail. I would welcome article suggestions from all readers of this column. Your suggestions could bring up new ideas that I haven't thought about before. Please send your request directly to me at either my e-mail address (dino.hunter@cox.net) or my Post Office Box, at PO Box 1171, Poway Ca, 92074.

Fred is having difficulty depicting theropod feet. I hope that I will be addressing his question in the way he needs.

First, the feet of dinosaurs need to be addressed. They are divided into different bones; the longest bones are the metatarsals (the heal, with the top being the ankle); phalanges (toe bones) and unguals (claws). Digit one (written as I) has one phalange, digit II has two, digit III has three, digit IV has four and digit V has five, with most of the toes ending in an ungual. The number of phalanges can vary, but for the most part this is the arrangement (figure 1).

In theropods, digits II and IV can be the same length or one longer than the other. Digit III (the middle toe) is the longest. Digit I (equal to your big toe and toe 'number' proceeding outward to digit V) is the smallest and can rest adjacent to or behind metatarsal II (as in birds). It can also be missing altogether (figure 2).

Theropods have at least three digits, with the majority of them having four. A perplexing problem about theropod feet is digit I's position. Was it on the side of metatarsal II or behind it? Perching birds have digit I on the lower back of digit II. This has been accomplished in two different ways. One the toe has twisted and flipped around backwards or two, has just flipped backwards. The question of the position of digit I in theropods is an age-old question.

In large theropods (I know for sure in allosaurids and tyrannosaurids) there is a small nearly oval indentation or dimple on the back of metatarsal II near its mid-shaft. I've asked paleontologist if this is where digit I would have been. They told me that it is a muscle scar. If this were true then a theropod that lacked digit I would also have this muscle scar. I've looked at metatarsal II in ornithimimids and it lacks the oval indentation. To me this means that it is not a muscle scar and is where metatarsal and digit I attached making it more bird-like in appearance. Unlike birds, digit I didn't reach the ground and couldn't be used for perching. Also unlike the dewclaw in a dog, the toe and claw had more mobility because of the tendon and muscle attachments thus it was more functional than the dewclaw of a dog.

Norell and Makovicky, (1997) described a new nearly complete *Velociraptor* specimen from Mongolia. The feet are nearly complete. The left foot (I have a cast of it) has digit IV missing the claw and the tip of the claw of digit III. Metatarsal I is lying on the side of metatarsal II at its mid-shaft. The first phalange and claw are lying parallel to metatarsal II. The right foot is nearly the same. Norell and Makovicky state that the toes have been disarticulated except for I-1, II-1 and III-1. I disagree; to me all these phalanges are displaced. I've asked Mark Norell about the placement and he is/was under the impression that all theropods had digit I on the side of digit II, as was stated in the article. I disagree with this because even though this specimen has digit I on the side of digit II doesn't mean all theropods had it the same way. Also phalange I along with the ungual are slightly misplaced, as are the toes on digit II and IV (figure 3). These toes are bunched (constricted) together with the large killer claw on digit II pulled back and off the ground. The best way to demonstrate the position of digit I is to lay your left or right hand flat. Relax and lay your thumb next to the palm. The nail (or claw in the dinosaur) would be parallel to the palm. Where as I believe if you flexed the tip of your thumb out, this would show how I'm interpreting the direction of the toe in the *Velociraptor* specimen lays in situ. With the unnatural placement of the toes, metatarsal I may also be misplaced and should have been on the back of metatarsal II.

What about the claw? Which way did it face? Toward the heel as a hallux in birds or away from it? I've also asked this question to several paleontologist and the majority of them answered that it faced away from the heel (in the same direction of the other claws), thus unlike a perching birds (figure 4). Ostrom (1970) has ungual I facing the heel in *Deinonychus*. This may be incorrect.

Next month I'll be going through the step cycle of an Emu (that is if I can get the San Diego Zoo to cooperate) as well as a closer look at the scales on the feet of birds.

Bibliography

Norell, M. A., and Makovicky P. J., 1997, Important features of the Dromaeosaur skeleton: information from a new specimen: American Museum Novitates, n. 3215, p. 1-17.

Ostrom, J. H., 1969, Osteology of *Deinonychus antirrhopus*, an Unusual Theropod from the Lower Cretaceous of Montana: Peabody Museum of Natural History, Yale University, Bulletin 30, p. 1-165.

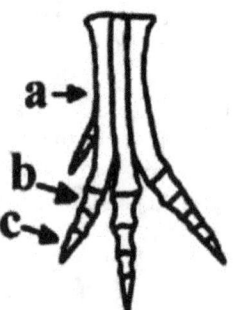

Figure 1). Generic theropod foot showing the different bones; A) metatarsals; B) phalanges and C) unguals).

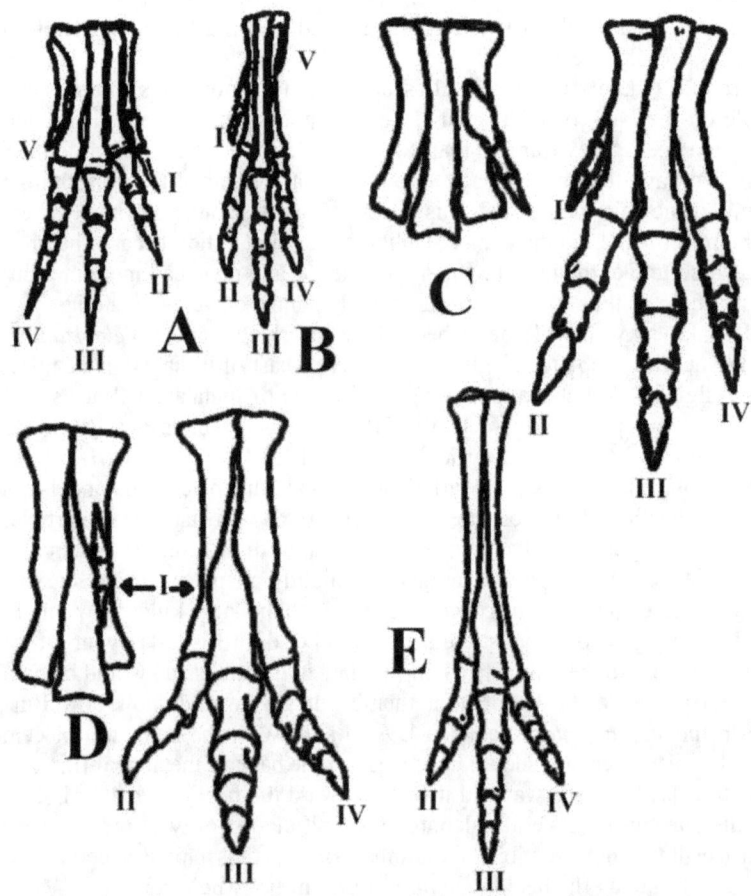

Figure 2). Feet of theropods; A) *Herrerasaurus*; B) *Syntarsus*; C) *Allosaurus*; D) *Tarbosaurus* and E) *Struthiomimus*.

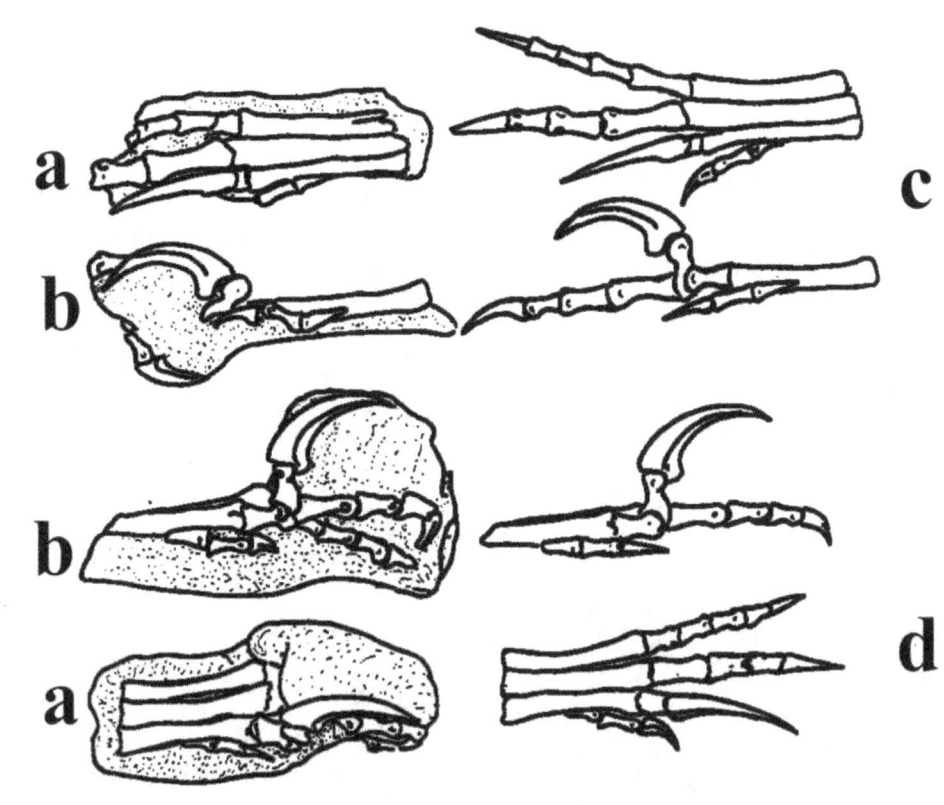

Figure 3). The feet of *Velociraptor*; A) lateral view of right and left foot; B) dorsal view of feet; C) close up the metatarsal I, phalange and ungual; D) corrected metatarsal I.

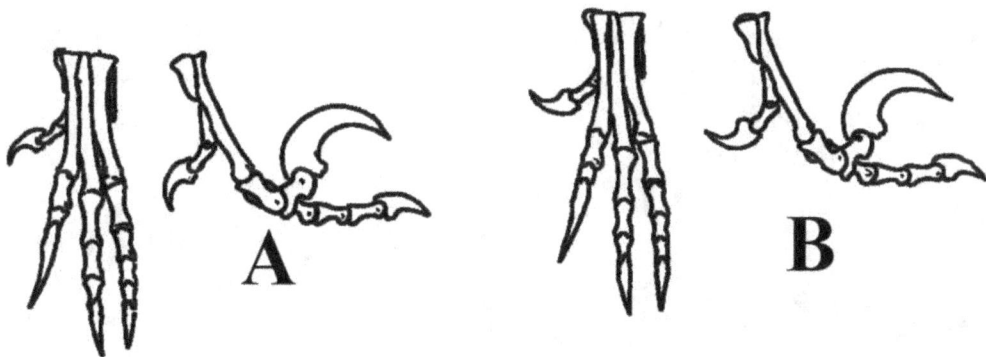

Figure 4). The foot of *Deinonychus* in; A) Ostrom's version of ungual I; B) corrected (?) version of ungual I.

For the Dinosaur Enthusiast and Collector

PreHiSTORIC TIMES

February / March 2001
No. 46

The PT Interview – Robert Walters Studio
Paleontologist Robert Bakker
PALEONEWS 2000 – Top Discoveries of Last Year

Dinosaur Art,
Collectibles, Books
& Model Reviews

U.S. $5.95 • Canada $6.95

Ford, T. L., 2000, How to Draw Dinosaurs. Walking with Emu's: Prehistoric Times, n. 44, p. 14-15.

Chapter 24

Walking with Emu's

First, allow me to make a few comments on the trials and tribulations of my research on the Emu's. I wasn't able to get a reply from the San Diego Zoo about the Emu at the bird show, so I went to the Wild Animal Park instead because I was told that their Emu's were sent there. Unfortunately it was closed when I got there. Besides, I learned from the Wild Animal Park information booth that the only Emu that they had was in their bird show. Fortunately there is an Ostrich farm (Wholesome Heritage Farm, family owned and operated, as stated by the sign) on the outskirts of the Wild Animal Park (which was also closed when I got there). But, luckily they had some Emu's as well as Ostriches in a fenced area close to the road. I was able to videotape and photograph them.

The foot of a bird can be used as a model in the illustrating or sculpting of dinosaurs, especially theropods, though there are differences. The ankle in birds is held at approximately 30 to 45°, as is theropods. Since birds lack the large balancing tail, the femur is held 20 to 40° from horizontal, and moves up and down. It is the lower leg which does all the movement. The tibiotarsus (lower leg) is held perpendicular to 30° from vertical in reference to the femur and when the metatarsals are vertical the tibiotarsus is at 45°. The metatarsals range from vertical to 30° (depending on whether the bird is standing or walking) (figure 1a). The balance point (or center of mass) for bird's is near the knee, which is why the legs are held they way they are. In theropods the balance point is at the pelvis and is offset by the tail. The femur is vertical or nearly so, and moves for and aft. The tibia/fibula is at about 40° to the femur and the metatarsals are 30 to 45° (Figure 1b).

The feet of birds have scales (a.k.a. scutellate tarsus) that cover the front of metatarsals. These scales are square or rectangular. The scales range in size from small oval scutes (like an Emu's or Peregrine falcon) to large single scales (as in some perching birds). Few birds besides the Emu have a fold of skin separating the toes from the scutellate tarsus.

The main difference between bird and theropod metatarsals is that in birds they are fused into a solid bone (tarsometatarsus). In the type skeleton of *Ceratosaurus* the metatarsals are fused due to pathology and is not a natural occurrence as originally thought by Marsh (I've done research on the pathology and will present it at a later date). There are no mummified feet of theropods so it is not known if they had small or large scutellate tarsus like emus or other birds. It is generally accepted that they did. Have others questioned this as I have? The metatarsals of theropods are separate bones and would have been similar to a sliding joint, passing by each other like a shock absorber, as the animal walked. If there were large square scutellate tarsus, they would have covered the front of the metatarsals. This might have hindered the movement of the metatarsals, similar to the way a *Stegosaurus* tail plates would lock the tail. Mummified hadrosaurs show that at least they didn't have a scutellate tarsus.

Birds have a round heel and oval toe pads (figure 2). Not all theropod tracks will show the heel pad, but will show the toe pads. This isn't to say they didn't all have a heel pad just that they were walking on their toes. When the foot is planted on the ground the pads will flatten out. The toes in cross-section will have a trapezoidal shape. When the foot is lifted the pads will become more rounded.

Picture the walking step. When the foot is firmly on the ground, the toes are spread apart. In most ground birds, the toes are more closely spaced. In perching birds, digits two and four are more spread out with digit one being the hallux or 'hind-toe'. This is one characteristic that is used in identifying Mesozoic bird tracks, which may go as far back as the Late Triassic. When the foot is lifted off the ground the heel is raised first and rolls off the toes. The toes will fold toward themselves and be at nearly a 45° angle at a slow pace to the metatarsals and 90° when running. I haven't watched Emu's running. I've only witnessed them walking, but I do have a marvelous book call Animals in Motion by Edward Muybridge. It shows an ostrich walking fast and running. At mid-stride the toes will be facing forward.

There is an experiment that you can due with your hand that will show how the toes work. If you place your fingers firmly on the ground and the back of your hand at an angle, this would be the natural resting plane of a theropod/ornithopod foot. If you bring your arm forward and roll off your fingers till you

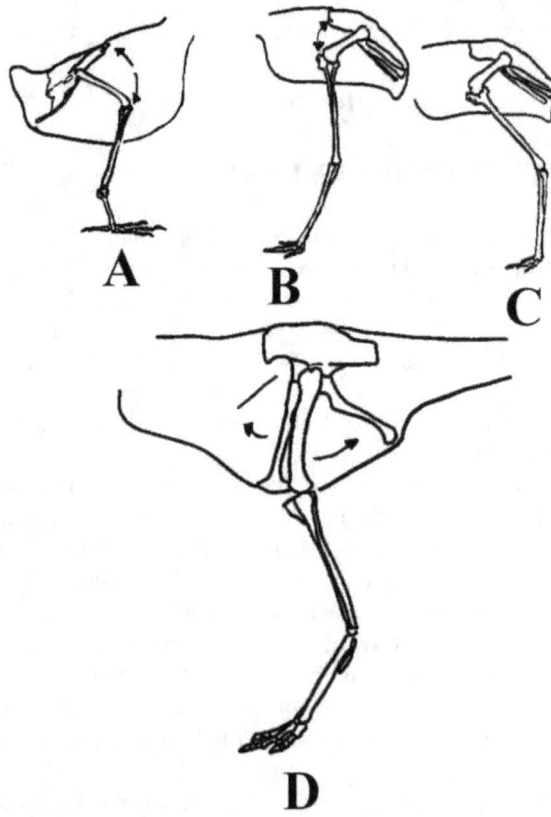

Figure 1). Legs of birds and dinosaur. A) Pigeon; B) Emu standing; C) Emu with vertical metatarsals (After Nobel S. Proctor and Patrick J. Lynch, 1993. Manual of Ornithology, Yale University Press); D) *Dromiceiomimus* (after Russell, 1972).

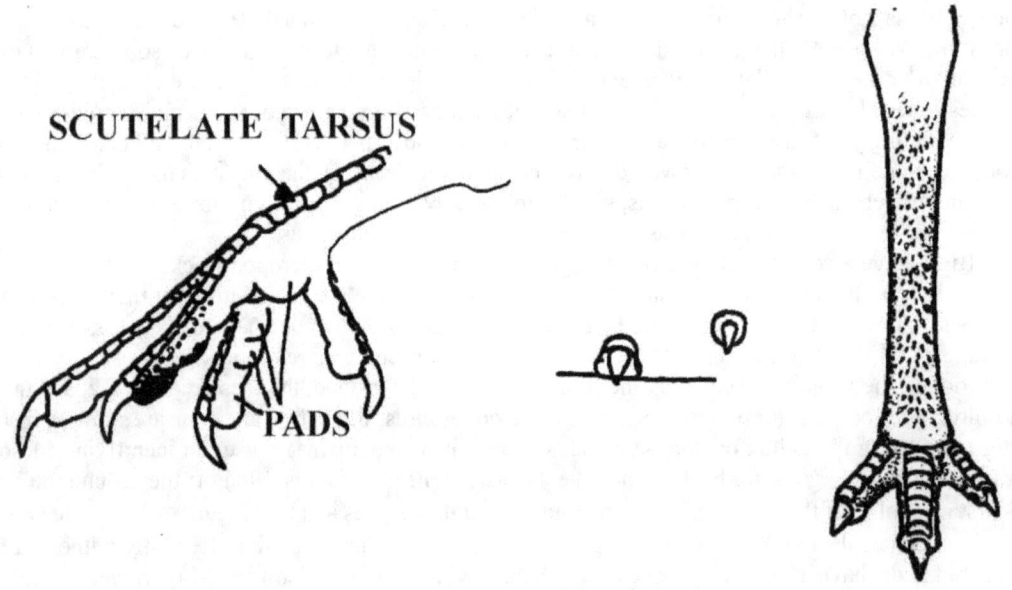

Figure 2). Foot of rock dove (after Proctor and Lynch, 1993), front view of toes, and foot of an Emu.

lift your hand up, you'll notice that your fingers will 'snap' close and be at about a 90° angle to your palm. This is the first step phase of the footfall. When walking slowly, the Emu lifts its foot off the ground when the metatarsals are slightly past vertical. This will vary due to the terrain and whether or not it is walking uphill or down.

When running, the Emu's foot is at nearly 45° past the vertical, and the toes will snap shut and be at more than a 90° angle toward the metatarsals. This is a good way to illustrate a fast moving animal.

During the step-phase, the metatarsals vary in its angle depending on how fast the animal is moving; from 45° when walking slowly to parallel to the ground when running. The tibotarsus moves from anywhere from 45° (the beginning of the step-phase) too nearly vertical (the mid to end step-phase, before the foot reaches the ground). In theropods this may have been the same or to a lesser degree. When the foot is brought back to the ground, the toes will automatically open and flare out. The toes will touch the ground first, followed by the heel. What I want to emphases is how the foot looks during the footfall. My illustrations will help show that (figure 4).

I would like to make a comment about *Amargasaurus*. I fault myself or not stating this in my previous articles about this magnificent beast, but the first spine coming off the atlas is singular, the rest are bifurcated or double. This means that nearly all those great illustrations that have appeared in PT are incorrect, and all artists should heed this.

Bibliography

Proctor N. S., and Lynch P. J., 1993, Manual of Ornithology, avian structure & function: Yale University Press, 340pp.

Figure 3). Step-phase of an Emu.

Figure 4). Step-phase of an Ostrich walking A); and running B) (Both after Muybridge).

164

For the Dinosaur Enthusiast and Collecto

PREHISTORIC TIMES

December 2000/January 2001 No.45

William Stout Part II

Jurassic Park in
Your Living Room

Giganotosaurus

U.S. $5.95 • Canada $6.95

165

Ford, T. L., 2000-2001, How to Draw Dinosaurs. The armored jigsaw puzzle of ankylosaurs: Prehistoric Times, n. 45, p. 14-15.

Chapter 25

The armored jigsaw puzzle of ankylosaurs

This issues spotlight dinosaur is the ankylosaur *Euoplocephalus*. *Euoplocephalus* is the most illustrated and best-understood ankylosaurid. This is due to nearly complete ankylosaurid specimen at the British Museum (Natural History). It was originally called *Scolosaurus cutleri* by Baron Franz Nopcsa in 1928. But is *Euoplocephalus* as well understood as we think?

The two people that stand out in the last 2 decades in depicting how ankylosaurs looked are George Olshevsky and Ken Carpenter. George had published an ankylosaur chart in 1979 (he is quite an artist) and some of his interpretations of ankylosaurs are still being used today. Ken's published two major papers on ankylosaur armor; one on the nodosaur *Sauropelta* and the other on *Euoplocephalus*. Both papers are widely used in the illustration of ankylosaurs. Ken used *Scolosaurus* (mummified skeleton lacking skull and most of the tail) as the main model in a paper he wrote in 1982 in which he described *Euoplocephalus* (numerous skulls, postcranial and pieces of armor). At the time both were thought to belong to the same genus, along with *Dyplosaurus* (known mainly from the pelvis to the end of the tail), who also thought to be the same animal.

I am currently researching ankylosaur armor and the traditional interpretation of how ankylosaurs looked will (I hope) be changed because of it. The only way to determine the position of armor is with an articulated specimen and they are few and far between. With my research I'm trying to establish terms for regions of armor as well as types of armor (figure 1). The regions are; the cranial (skull), cervical (neck), pectoral (shoulder), thoracic (back), pelvic (pelvis) and caudal (tail). Also starting on the top of the back (the armor is in sets of two) the mid-line armor is the medial, the next is the primary, secondary then the tertiary (which can be scutes [also called osteoderms] not set in rows over the thoracic area).

Ankylosaurid ankylosaurs have two sets of cervical half rings with 3 sets of scutes each. These scutes can be attached to a solid bony ring of bone (which wouldn't be seen in a living animal because it would have been hidden under the skin). The type specimen of *Euoplocephalus* (which might be a juvenile or just a small animal) has scutes that are large and touch each other. They have a large ridge down the middle and are nearly oval in shape and are connected to a thin, bony ring. In other individuals, the scutes are smaller with a gap between them, and in some cases the scutes are missing altogether (figure 2). Scutes have been known to have anything from a nearly smooth finish to very vascular surface. This may be due to ontogeny, sexual dimorphism or individual variation.

Carpenter pointed out that *Scolosaurus* did not have cervical rings like that of *Euoplocephalus* (Figure 3). This may be because of *Scolosaurus'* mummification and the rings may be hidden under the skin. *Scolosaurus* has the typical two 'bands' of cervical armor, with the scutes separated by irregular small flat ossicles (skin). The first band has two medial rectangular conjoined plates with weakly developed ridges (unlike *Euoplocephalus*). The primary scutes have oblong bases and a ridge that points posteriorly. The second band has two very large, nearly flat oval medial scutes with a large gap over the middle of the neck. The lower edge of the body is missing, thus the secondary set of scutes may be missing. Between these bands are small irregular flat ossicles that are smaller than the ossicles on the cervical bands. The osteoderms of *Scolosaurus* have been eroded and were certainly larger in life.

The body and pelvis have fleshy transverse bands with osteoderms, which are not connected to bony bands like the cervical bands. There are four transverse bands covering the body, two over the shoulder and two over the back. All the scutes have small polygonal ossicles around the bases and in between the scutes with smaller polygonal ossicle separating the bands. Over the shoulder the medial scutes are the tallest with rounded bases. The medial scutes are set in a paired row down the middle of the back with the scutes getting smaller and shorter toward the pelvis. The primary scutes have oval bases, are triangle in lateral view with the ridge's peak toward the posterior end. The secondary scutes are of a similar shape with a much taller ridge point. One side of the body is more complete than the other, so the exact number of scute rows that run along the body is unknown. The front left leg of *Scolosaurus* has a cluster of osteoderms on the lower portion (from the elbow down). They have oval to nearly square bases with low ridges near the midline.

As noted before, there are two bands of osteoderms covering the back of *Scolosaurus*. These bands are very similar to the previously mentioned pectoral bands, only that the scutes are smaller. There are fragmentary secondary scutes found along the sides. The osteoderms on the left half of the pelvis were removed to show the ilia and sacral ribs, but the right half has the pelvic osteoderms preserved. The sacral region has three sets of un-fused bands of osteoderms. These bands have oval scutes that have short nearly horizontal rounded peaks. Near the posterior edge of the right side of the pelvis there appears to be a cluster of four oblong, posteriorly peaked scutes.

Only the first half of the tail of *Scolosaurus* is known with four bands preserved (similar to the thoracic bands) with a pair of triangular laterally compressed medial scutes. The fourth band has the largest medial scute (which was interpreted as the club at the end of a short tail with two large spikes). Around the base of the scutes are asymmetrical mosaic ossicles with smaller asymmetrical ossicles on the bands. Like the body bands, there are smaller irregular ossicles in between the bands.

Dyoplosaurus acutosquameus (another small ankylosaurid) has a more complete tail preserved. There are several posteriorly pointed triangular, laterally compressed, primary caudal scutes on the side of the tail. These scutes stopped at the mid section of the tail where the long ossified tendons of the tail begin. These probably ran down the length of the tail. The distal end of the tail has a typical ankylosaur tail club.

The armor of "North American" Ankylosaurids differs from the "Asian" forms. The "Asian" forms are more heavily armored. Unfortunately only a few of the articulated specimens from Mongolia have ever been collected. I am trying to keep in touch with the institutions that have collected them to study the placement of the armor.

The armor of *Ankylosaurus* is totally different from *Scolosaurus*, *Euoplocephalus* and the "Asian" forms. I hope in the (?) future to describe the armor of *Ankylosaurus* and show its difference to the other ankylosaurs. (**Editors note: Arbour & Currie, 2013, has shown that *Scolosaurus*, *Euoplocephalus* and *Dyoplosaurus* are valid genera. This means we know what *Scolosaurus* looked like, but not the other two**).

Appendix

Scolosaurus, *Euoplocephalus* and *Dyplosaurus* are known from the Dinosaur Provencal Park, Judith River Group of the Dinosaur Park Formation of the Late Campanian, Late Cretaceous.

Bibliography

Arbour, V. M., and Currie, P. J., 2013, *Euoplocephalus tutus* and the diversity of ankylosaurid dinosaurs in the Late Cretaceous of Alberta, Canada, and Montana, U.S.A: Public Library of Science (PLOS), One, v. 8, n. 5, 39 pp.

Carpenter, K., 1982, Skeletal and dermal armor reconstruction of *Euoplocephalus tutus* (Ornithischia: Ankylosauridae) from the Late Cretaceous Oldman Formation of Alberta: Canadian Journal of Earth Science, v. 19, p. 689-697.

Carpenter, K., 1984, Skeletal reconstruction and life restoration of *Sauropelta* (Ankylosauria: Nodosauridae) from the Cretaceous of North America: Canadian Journal of Earth Science, v. 21, p. 1491-1498.

Lambe, L. M., 1902, New genera and species from the Belly River Series (Mid-Cretaceous): Contributions to Canadian Palaeontology, v. 3. Geological Survey Canada, p. 22-81.

Nopcsa, F., 1928, Palaeontological notes on Reptiles: Geologica Hungarica, Series Palaeontologica, tomus, 1, -Pasc. 1, p. 1-84.

Parks, W. A., 1924, *Dyoplosaurus acutosquameus*, a new genus and species of armored dinosaur. Notes on a skeleton of *Prosaurolophus maximus*: Univeristy of Toronto Studies, Geological series, n. 18, 35pp.

Penkalski, P., 2001, Variation in specimens referred to *Euoplocephalus tutus*: In: The Armored Dinosaurs, edited by Carpenter, K., Indiana University Press, p. 261-298.

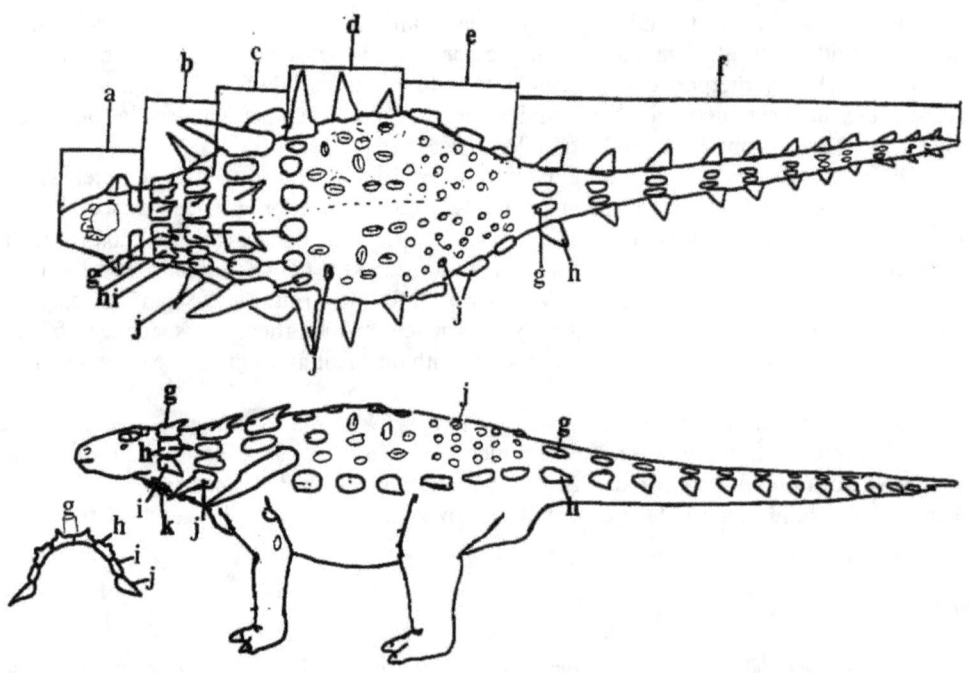

Figure 1). Illustration showing the different areas of armor. (a) cranial, (b) cervical, (c) pectoral, (d) thoracic, (e) pelvic, (f) caudal regions, (g) medial, (h) primary; (i) secondary, (j) tertiary and (k) ventral osteoderms.

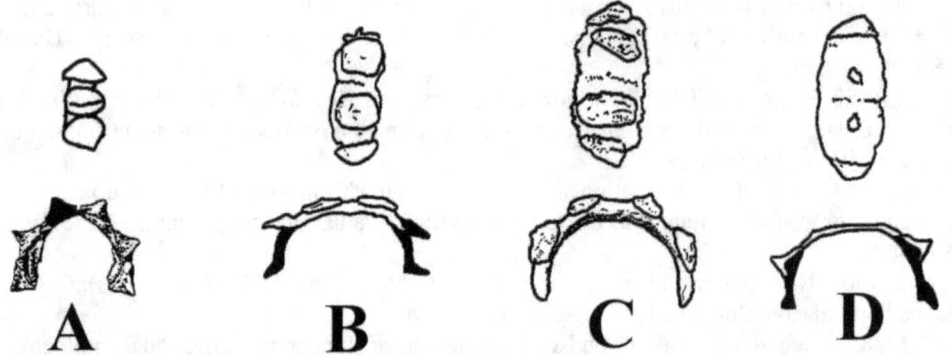

Figure 2A) The cervical ring of the type of *Euoplocephalus*, NMC 0210 from Lambe (1902); B-D) different rings showing the different states of cervical osteoderms on the 'bony' ring (after Penkalski, 2001); B) USNM 7943; C) AMNH 5337; D) BMNH R5161 .

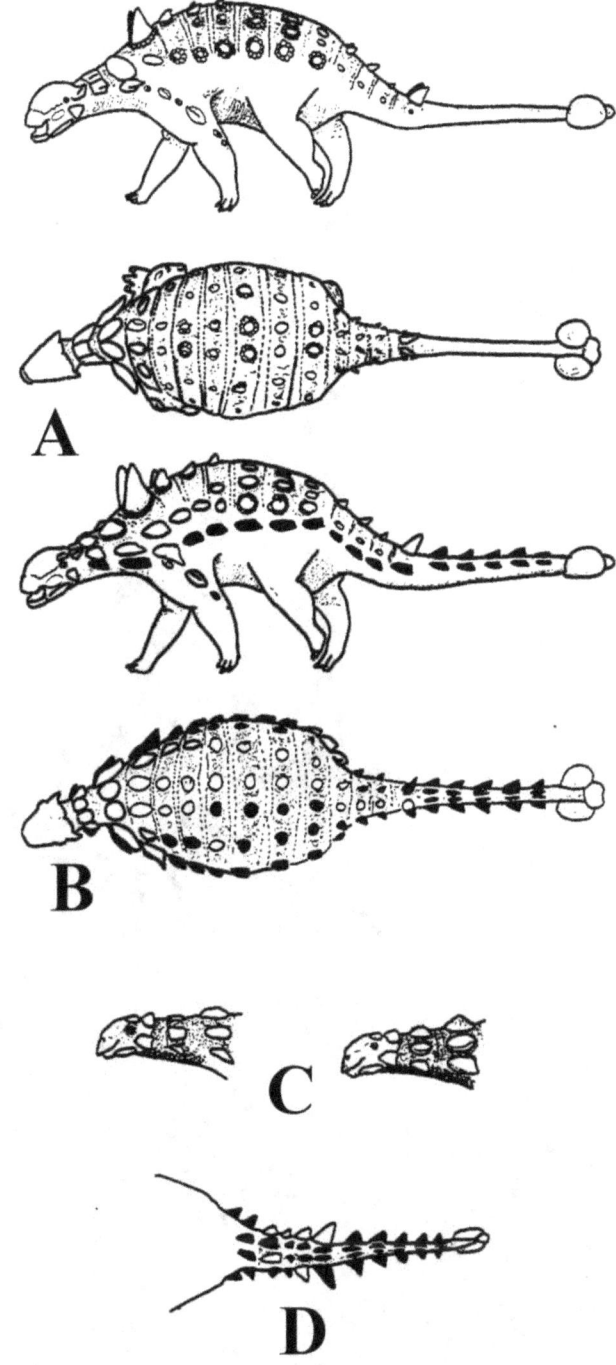

Figure 3A). *Scolosaurus* after Ken Carpenter; B) modified *Scolosaurus*; C) *Euoplocephalus* cervical region; D) *Dyplosaurus* tail.

PREHISTORIC TIMES

NO. 44   OCT/NOV 2000

WILLIAM STOUT
THE NEW DINOSAURS

INTERVIEW
WITH
PHIL CURRIE

LUIS REY
RUFFLES SOME
FEATHERS

U.S. $5.95 • Canada $6.95

0 56698 94980 0

10

Ford, T. L., 2001, How to Draw Dinosaurs. Anatomy 101: Prehistoric Times, n. 46, p. 14-15.

Chapter 26

Anatomy 101

My topic this time is something that I should have discussed a long time ago; an article on the names of the different bones of the skeleton so the reader will have a 'map' to go by. I won't go into all the different little bones or the braincase or the names of all the struts and creases of vertebrae. Instead I will focus on the bones that will best help the artist and layperson alike and I will use *Allosaurus* as the model. So here is Anatomy 101.

First, I'll list the names of the 'holes' (or fenestrae) in the head of dinosaurs (figure 1). From the side (lateral view) you can see the external naris (nose); behind that is the antorbital fenestra (not all dinosaurs have this fenestra); the orbital fenestra (eye); behind that is lateral temporal fenestra and above that is the upper temporal fenestra. The upper temporal fenestra is where the jaw muscles come out of and attach to the parietal. What function the fenestra had is still unknown. Whether or not the head showed the fenestra as a relief or was 'smooth' is up to the artist (**Editors note: I wrote about this subject in a later article**). Larry Witmer and his group are doing research to answer this often asked question.

Nearly all the bones of the skull are paired (figure 2). Starting from the front of the skull there is the premaxilla, then the maxilla. These are the tooth bearing bones of the upper skull. It is important to find out how many teeth (and the shape of the teeth) each bone has. Only a few ornithischians have premaxillary teeth. The premaxilla and pre-dentary of ornithischians (the bone on the tip of the lower jaw) had a keratin covering making a horny beak. Some theropods lack teeth altogether (ornithmimosaurs, oviraptorids to name a few). Behind the maxilla is the jugal, then the quadratojugal. Lying nearly perpendicular to the quadratojugal is the quadrate. It is the lower posterior extension of the quadrate that forms the upper joint of the jaw. Around the middle of the quadratojugal/quadrate is where the ear would be, and did not sit high on the skull.

Back to the front of the skull and moving up and back are the nasals, lachrymal, the frontal and then the parietal. Each of these bones can have horns, bumps or 'sail'. I'll be going into some of these cranial 'displays' next issue for theropods and the issue after that for the ceratopians. On the upper side of the skull and separating the lateral temporal fenestra from the temporal fenestra, is the postorbital then the squamosal. The occipital condyle lies on the back of the skull and this bone (a ball for the socket of the atlas/axis cervical vertebrae) connects to the neck. The angle of this bone in relation to the neck will determine how the head is held. For example, in *Allosaurus* the occipital condyle is horizontal to the neck (and the head would be parallel to the ground) while in *Tyrannosaurus rex* the condyle is at a 45° angle to the neck (and the head would tilt downward).

The lower jaw has only a few bones that we need to worry about. There is the dentary (which is the only toothed part); behind and dorsal to that is the surangular (which is where the jaw muscles attach to) then the articular (which is the lower bone for the jaw joint). The lower bone is the angular.

On to the skeleton (figure 3). The vertebral column break down is as follows; the cervical (neck), dorsals (back), sacral/sacrum (pelvis), and caudal (tail). The vertebra itself has the centra (lower portion) the neural arch (section between the centra and neural spine) and the neural spine (figure 4). An unfused neural arch and centra are indications of an immature animal. The wing-like extension of the vertebra is the transverse process and the tip of the process is the diapophysis. The diapophysis is where the upper part of the rib will attach to the vertebra. The angle of the transverse process and the arch of the rib will determine how wide the body is.

The cervical vertebrae have paired cervical ribs. These ribs lie parallel with the bottom of the centra. They did not extend ventrally like normal ribs. The 'rod' like bones on the ventral side of the caudal centra is the chevrons. They lie between adjoining caudal centra.

The shoulder girdle or pectoral girdle (figure 5) consists of the scapula, coracoid, sternum (which forms the 'chest') and in some theropods a furcula or wish bone (the number of theropods found that had a furcula is increasing). The arm includes the humerus, ulna, radius, several smaller wrist bones, metacarpals, phalanges (fingers) and unguals (claws). Some dinosaurs have ventral ribs (or gastralia).

The pelvic girdle or hip is made up of the ilium (large upper bone), which the sacral vertebrae attach, the pubis (front lower bone) and the ischium (back lower bone, figure 6). In some dinosaurs the front of the ilia will flare outward and many artist fail to take this flare into account. The leg consists of the femur, tibia, fibula, some ankle bones, metatarsals, phalanges (toe bones) and unguals (claws).

This concludes Anatomy 101 and there will not be a quiz.

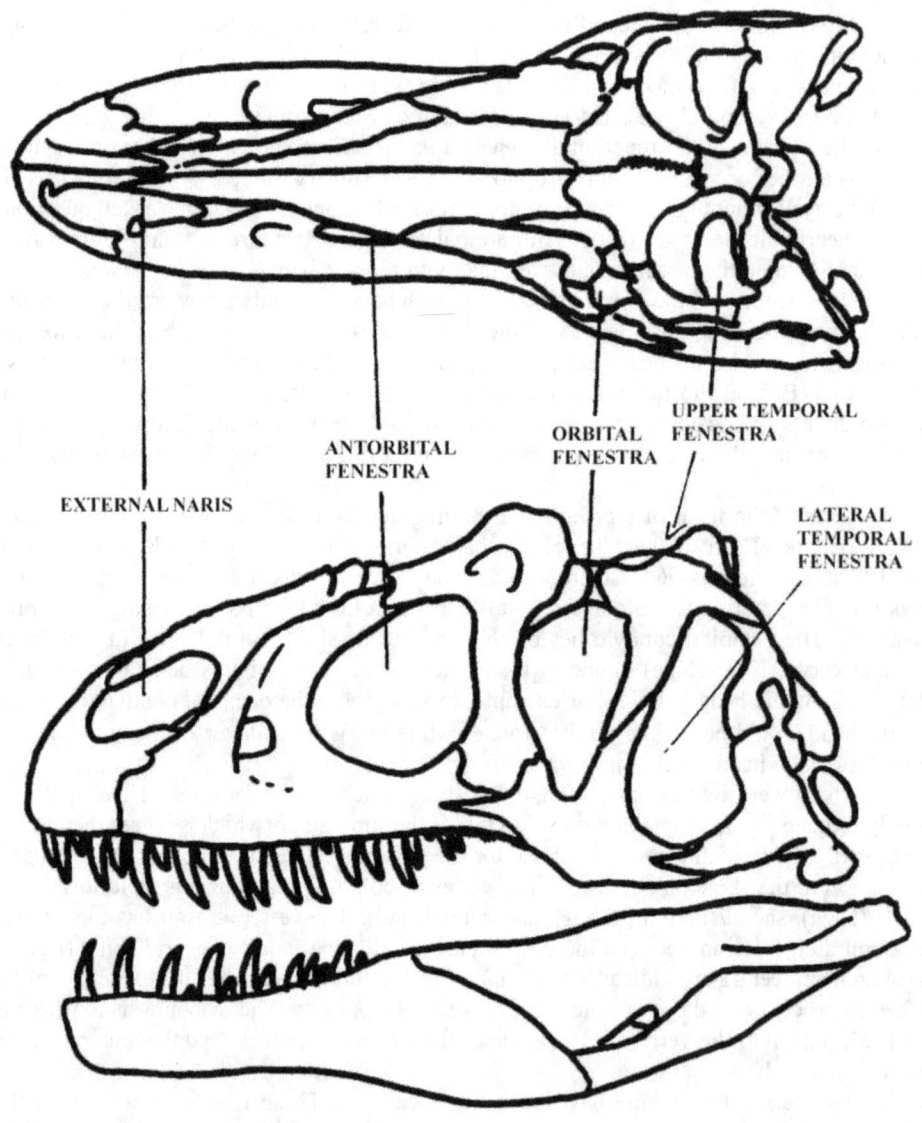

Figure 1) The fenestra.

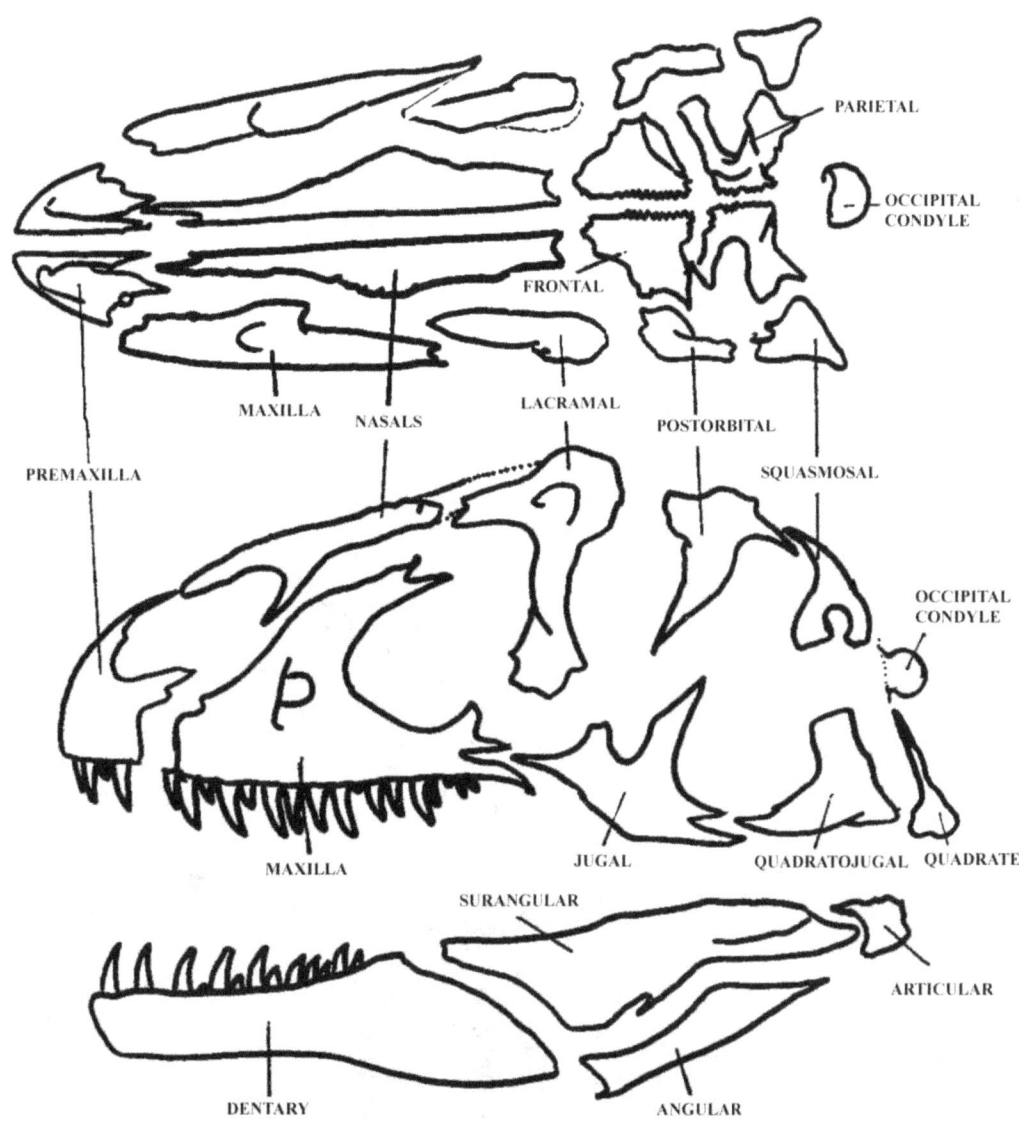

Figure 2) The Skull.

173

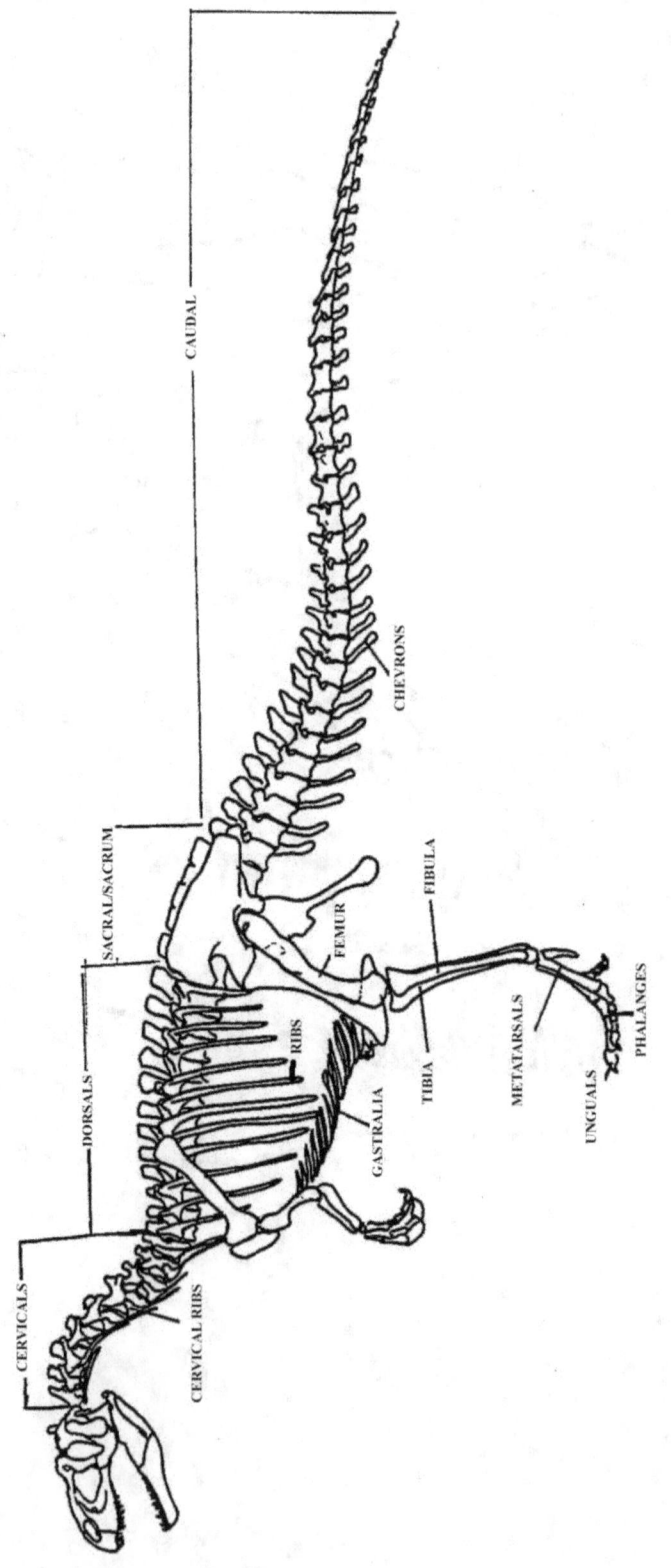

Figure 3) The skeleton.

174

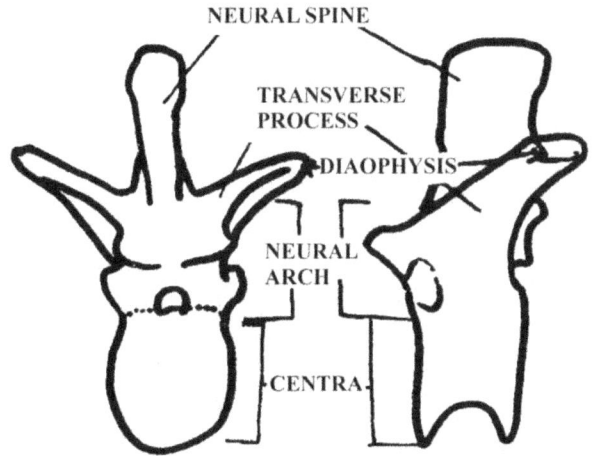

Figure 4) The vertebra.

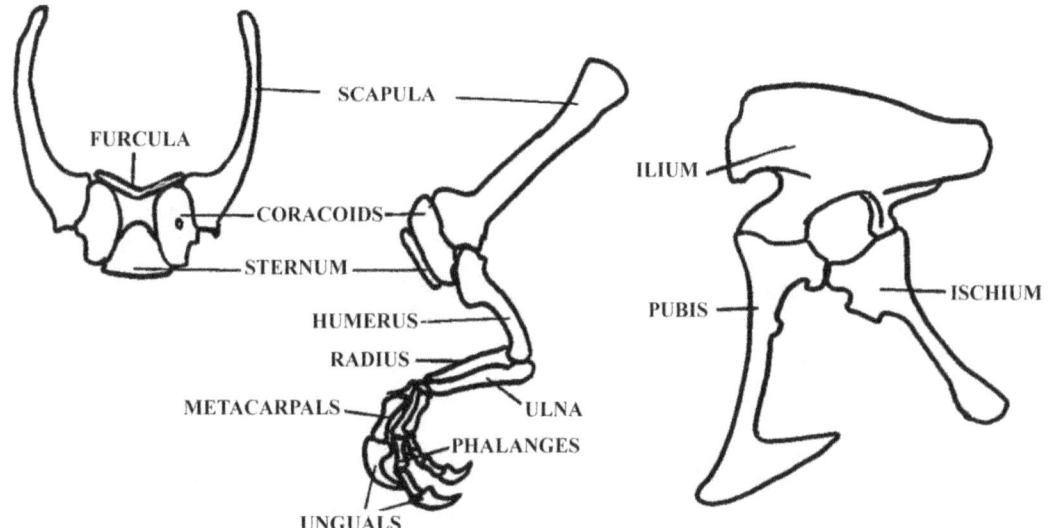

Figure 5) The pectoral girdle.

Figure 6) The pelvic girdle.

www.ingramcontent.com/pod-product-compliance
Lightning Source LLC
Chambersburg PA
CBHW080809180526

45168CB00006B/2379